Biofortification in Rice

NIPA® GENX ELECTRONIC RESOURCES & SOLUTIONS P. LTD.
New Delhi-110 034

About the Authors

Dr. Krishnendu Chattopadhyay did his MSc. Genetics from Indian Agricultural Research Institute, New Delhi and completed Ph.D. Genetics from Bidhan Chandra Krishi Viswavidyalaya, West Bengal. He join Agricultural Research Service in 1999 and at present he is working as Principal Scientist at ICAR-Central Rice Research Institute, Cuttack. During his 25 years of intensive research work, he has developed and released 20 varieties in various crops and published more than 60 research papers in national and international reputed journals. Among these varieties CR Dhan 310 is the first high protein rice variety of India. It is landmark biofortified variety emerging as mega rice variety with commercial importance. High zinc variety CR Dhan 315, dedicated to nation by the Hon'ble Prime Minister and 'high protein Swarna', Swarnanjali are also exhibiting potentiality to replace the popular varieties ensuring better nutrition. He is also lead developer of multiple stress tolerant rice varieties for coastal saline areas and registered germplasm for abiotic stress tolerance and for nutritional quality. He has contributed in genetics of high grain protein and abiotic stress tolerance by identifying robust QTLs. As a team leader he was awarded 'Nanaji Deshmukh ICAR Award for Inter-disciplinary team research- 2018' in the field of biofortification in rice. He was also awarded Fellow of Indian Society of Genetics and Plant Breeding and Fellow of Association of Rice Research Workers.

Dr. Koushik Chakraborty, Senior Scientist (Plant Physiology) is presently working at ICAR-Central Rice Research Institute, Cuttack. Dr. Chakraborty has done his post-graduation and PhD in Plant Physiology from Indian Agricultural Research Institute, New Delhi. He carried out post-doctoral research at the University of Tasmania, Hobart, Australia in 2014 under Australian Govt. sponsored Endeavour Research Fellowship. His primary research area includes abiotic stress physiology (salinity, submergence and stagnant flooding) and grain nutritional quality of rice. His works elucidated the membrane ion transport mechanism and role of different transporters/ion-channels/pumps in stage- and tissue-specific salt tolerance strategies of rice and mustard. He is the recipient of prestigious Lal Bahadur Shastri Young Scientist Award from ICAR, Govt. of India (2022), Associateship of National Academy of Agricultural Sciences (2021), Associate Fellow of West Bengal Academy of Sciences (2020), Membership of National Academy of Sciences, India (2023), Norman E. Borlaug International Agricultural Science and Technology Fellowship Program by United States Department of Agriculture, Foreign Agricultural Service (USDA-FAS) (2014),

'Endeavour Research Fellowship' by Australian Government Department of Education, Govt. of Australia (2013), ISCA Young Scientist Award by The Indian Science Congress Association, Department of Science & Technology, Govt. of India (2013). He has published more than 55 research article in SCI indexed journals, identified 15 unique genetic stocks of rice and groundnut and also associated with development of 6 rice varieties. Besides, Dr. Chakraborty has also guided 07 M.Sc. and 02 Ph.D. scholars for their thesis/dissertation work.

Dr. Torit Baran Bagchi, ARS is a Scientist of ICAR-Central Rice Research Institute in the department of Crop Physiology and Biochemistry. He received "Nanaji Deshmikh ICAR Award for Outstanding Interdisciplinary Team Research in Agriculture and Allied Sciences" in 2018 as associate for the development of first high protein rice variety CR Dhan 310 and another 3 biofortified rice varieties. He published more than 35 research article in reputed national and international journals. He also received one patent (Patent No.383679) regarding "A multiuse composition for bio-control of plant pathogen infestation and growth enhancement". Besides, he is the author of many book chapters, research bulletins and scientific popular articles, published in different scientific magazines and ICAR publishing units. He guided five MSc students and running three external funded projects of DBT, RKVY and ICAR National fund. His research work is mainly associated with development of bio-fortified rice, value added products from rice flour and assessment of rice grains for higher antioxidants, minerals, proteins and vitamins.

Biofortification in Rice
Present Status and Future Prospects

Krishnendu Chattopadhyay
Principal Scientist
ICAR-Central Rice Research Institute
Cuttack-753006, Odisha, India

Koushik Chakraborty
Senior Scientist (Plant Physiology)
ICAR-Central Rice Research Institute
Cuttack-753006, Odisha, India

Torit Baran Bagchi
ARS, Senior Scientist
ICAR-Central Rice Research Institute
Cuttack, Odisha, India

NIPA® GENX ELECTRONIC RESOURCES & SOLUTIONS P. LTD.
New Delhi-110 034

NIPA® GENX ELECTRONIC RESOURCES & SOLUTIONS P. LTD.

101,103, Vikas Surya Plaza, CU Block
L.S.C. Market, Pitam Pura, New Delhi-110 034
Ph : +91-11-43860225, Mob.: +91 9717133558, 9540816132
E-mail: newindiapublishingagency@gmail.com
Website: www.nipaersources.com

Print ISBN: 978-93-58872-30-9
ebook ISBN: 978-93-588-78-27-1

Composed and Designed by NIPA®.

Preface

As one of the fastest-growing global economies in the world India faces contradiction with having an estimated 14% of the population as undernourished. Furthermore, the malnutrition status and immunity of vulnerable communities in India has been diminished due to COVID-19 pandemic. India is self-sufficient in food grain production. Policy approaches that tackle hunger through supplements, food provision via public distribution systems. The rate of improvement in nutritional status among a large section of the population has not kept pace with India's significant gains in economic prosperity as well as increasing productivity during recent decades in spite of governmental plan to tackle hunger. Therefore, policy support has been gradually sifting from eradication of hunger to diminution of hidden hunger. Biofortification is one of the cost effective and environmentally and socially acceptable and ecologically friendly approach to enhance micronutrients, protein, vitamins and other important nutrition traits in edible food crops. Rice is the staple food of more than half of the Indian population. Therefore, biofortification in rice has substantial potentiality to enhance the nutritional status of the rice consumers. The Government of India has already taken an important step by providing policy and research support in linking agriculture and nutrition with biofortification. Prime Minister Narendra Modi has given a strong advocacy to staple crop biofortification as a natural and cost-effective solution to lessen malnutrition. On World Food Day in 2020, the Hon'ble Prime Minister dedicated to the nation 17 recently-developed biofortified varieties of eight indigenous and traditional crops, including rice, lacking key micronutrients that are essential for good health. He said, "We are taking an important step in our fight against malnutrition as we are now encouraging the production of crops which are rich in nutritional substances like protein, iron, zinc etc.".

This book is being published at a moment when the various aspect of biofortification is being considered at policy level and focused as an important tool to eradicate malnutrition and hidden hunger in India. Throughout the book all important nutritional traits of rice such as grain protein, essential amino acids, Fe, Zn, other minerals, Vitamins, antioxidants and also anti-nutritional factors are thoroughly explained. It covers most important aspects of nutritional traits of rice such as its estimation procedure, biochemical, molecular and physiological basis of qualitative and quantitative variation as well as its elevation in high yielding popular varieties.

Chapter 1 introduces the requirements and aspects of biofortification in rice in the present context of malnutrition among a large section of Indian population. Chapter 2 deals with the N metabolism and biochemical basis of assimilation of storage protein in developing grains. It also reviews the quantitative trait loci and genes responsible for high grain protein and amino acid content. Chapter 3 freshen readers on estimation procedure of grain protein, protein fraction and amino acid content as well as diversity among rice germplasm for those traits and breeding for high protein and amino acid content. Chapter 4 tells about nutritionally important micronutrients (Fe, Zn) and vitamin (A), their biosynthesis, metabolic regulation in rice and molecular basis of their enhancement in grain. In continuation, Chapter 5 narrates diversity of micronutrients in grains, its genetic basis including reported QTLs and improvement through breeding techniques. Chapter 6 elaborates role and scope of other nutritional and anti-nutritional factors in biofortification such as folate, selenium, vitamins, phytic acids as well as about glycaemic index and antioxidant activities influencing human health. Chapter 7 reviews the details of modern tools such as transgenic approach and gene editing in biofortificantion for important nutritional traits in rice. Finally, Chapter 8 depicts the constrains and future endeavour including researchable issues in biofortification in rice.

This book is textbook for students who want to want to develop basic understanding in nutritional traits in the backdrop of biofortification programme. This book has made detail review on recent research work conducted not only in India but in World for understanding biochemical and molecular basis of important nutritional traits which will be really helpful for researchers for designing future research initiatives in biofortification in rice. The recent varietal improvement in rice for nutritional traits and biofortified rice varieties especially in India and also in abroad are elaborated in details in this book. This is also will be useful for policy makers for fabricating governmental programme to eradicate malnutrition through naturally available nutritional diet and supplements.

Krishnendu Chattopadhyay
Koushik Chakraborty
Torit Baran Bagchi

Contents

1

Importance of Biofortification in Rice with Special Reference to India

Introduction

Malnutrition remains a pervasive global challenge, affecting millions of individuals across the world, particularly in low- and middle-income countries. Despite significant advancements in agriculture and food production, a large segment of the population still lacks access to essential nutrients, leading to severe health consequences and perpetuating cycles of poverty and underdevelopment (Müller and Krawinkel, 2005). Malnutrition, defined as the inadequate intake of essential nutrients required for optimal health and development, encompasses both undernutrition and over-nutrition (WHO, 2023). While undernutrition, characterized by deficiencies in energy, protein, vitamins, and minerals, remains a significant concern, over-nutrition, manifesting as obesity and diet-related non-communicable diseases, is also on the rise, particularly in urban settings (Wells *et al.*, 2020). The consequences of malnutrition are far-reaching, impacting physical growth, cognitive development, immune function, and overall well-being, with long-term implications for productivity, economic prosperity, and societal stability (Saunders and Smith, 2010).

Rice, a staple food for more than half of the world's population, plays a central role in global food security and nutrition. However, conventional rice varieties often lack sufficient levels of key nutrients, such as iron, zinc, vitamin A, and certain amino acids, leaving populations reliant on rice-based diets vulnerable to micronutrient deficiencies and associated health risks (Peña-Rosas and Mithra, 2019). Iron deficiency anaemia, zinc deficiency, and vitamin A deficiency, are among the most prevalent forms of malnutrition worldwide, particularly affecting women and children in resource-constrained settings where rice consumption is high (Dipti *et al.*, 2012).

Biofortification, defined as the process of enhancing the nutritional content of crops through conventional breeding, genetic engineering, or agronomic practices, offers a sustainable and cost-effective approach to addressing malnutrition (Talsma and Pachón, 2017). By targeting key nutrients directly in staple crops like rice, biofortification aims to improve dietary quality and alleviate micronutrient deficiencies at the population level, particularly among vulnerable groups with limited access to diverse and nutritious foods (Peña-Rosas and Mithra, 2019). Over the past two decades, significant progress has been made in developing biofortified rice varieties with enhanced levels of essential nutrients. Iron-biofortified rice, commonly referred to as "iron rice," has been successfully developed using conventional breeding techniques, resulting in varieties with significantly higher iron content than conventional rice varieties (Trijatmiko *et al.*, 2016). Similarly, zinc-biofortified rice varieties have been developed to address zinc deficiency, particularly prevalent in regions where rice is the primary staple food (Rao *et al.*, 2020). In addition to enhancing the content of individual nutrients, efforts have been made to develop rice varieties with multiple biofortified traits, known as "nutrient-rich rice." Biofortified rice varieties offer several benefits for nutritional security (Das *et al.*, 2023). By increasing the availability of essential nutrients in rice, biofortification helps to combat micronutrient deficiencies, such as iron-deficiency anaemia and zinc deficiency (Majumder *et al.*, 2019). Improved access to biofortified rice can enhance the overall nutritional intake of vulnerable populations, leading to better health outcomes, particularly among children and pregnant women. These varieties aim to provide a comprehensive solution to malnutrition by addressing multiple nutrient deficiencies simultaneously, thereby maximizing health impact and cost-effectiveness (Ofori *et al.*, 2022).

Biofortification offers a sustainable approach to addressing malnutrition, as it integrates nutritional enhancement directly into the food supply. Unlike supplementation or fortification, which may require ongoing interventions, biofortified crops can provide a continuous source of essential nutrients through regular consumption (Bouis and Saltzman, 2017). By promoting the cultivation and consumption of biofortified rice, countries can achieve long-term improvements in nutritional security. Biofortification efforts in rice have primarily targeted key nutrients such as iron, zinc, and provitamin A (beta-carotene) (Paine *et al.*, 2005). Various strategies, including conventional breeding, marker-assisted selection, and genetic engineering, have been employed to enhance the nutritional content of rice grains. Several biofortified rice varieties with elevated levels of micronutrients have been developed and released for cultivation in different regions (Garg *et al.*, 2018). Biofortified rice varieties exhibit significantly higher levels of iron, zinc, and provitamin

A compared to conventional rice varieties (Majumder *et al*., 2019). Iron and zinc play crucial roles in immune function, cognitive development, and overall health, while provitamin A is essential for vision and immune function (Bouis, 2003). By increasing the bioavailability of these nutrients in rice, biofortification has the potential to address micronutrient deficiencies and improve the nutritional status of vulnerable populations (Ofori *et al*., 2022). The widespread cultivation and consumption of biofortified rice have the potential to significantly reduce the prevalence of malnutrition, especially in rice-dependent communities. By integrating biofortified rice into existing agricultural systems, governments and organizations can promote sustainable solutions to malnutrition. Additionally, educational initiatives aimed at raising awareness about the nutritional benefits of biofortified rice can further enhance its impact on public health (Sandhu *et al*., 2023).

The significance of Biofortification in Rice, Particularly in the Context of India

India, with a substantial portion of its population relying on rice as a staple food, faces significant challenges related to malnutrition and hidden hunger (Bouis and Saltzman 2017). Biofortification of rice holds immense potential to address the nutritional challenges faced by India, offering a sustainable and scalable solution to improve the health and well-being of its population on the following aspects.

1. *Nutritional Deficiency:* Many regions in India, especially among economically disadvantaged communities, suffer from widespread nutritional deficiencies (Venkatesh *et al*., 2021). Biofortified rice offers a sustainable solution to address these deficiencies by enhancing the nutrient content of a staple food that is widely consumed across the country.
2. *Combatting Micronutrient Deficiencies:* Biofortification targets specific nutrients such as iron, zinc, and vitamin A, which are commonly deficient in diets heavily reliant on rice. These micronutrients are crucial for overall health, immune function, and cognitive development, particularly in vulnerable populations like children and pregnant women (Sood *et al*., 2023).
3. *Addressing Public Health Challenges*: Malnutrition, including both undernutrition and hidden hunger, poses significant public health challenges in India. Biofortified rice has the potential to mitigate these challenges by providing a more nutrient-rich staple food, thereby improving the overall health and well-being of the population (Yadava *et al*., 2018).

4. *Accessibility and Affordability*: Biofortified rice can be developed to be cost-effective and easily accessible to the masses, especially in rural and remote areas where nutritional interventions are often limited. This makes it a practical solution for addressing malnutrition among India's diverse population.
5. *Long-Term Impact*: Unlike short-term interventions, such as food supplementation programs, biofortification offers a sustainable long-term solution to address malnutrition. By enhancing the nutritional content of rice through genetic and agronomic approaches, biofortification can have a lasting impact on improving the health outcomes of future generations.
6. *Contribution to Food Security*: Biofortified rice not only improves the nutritional quality of the diet but also contributes to food security by ensuring a more balanced and diverse diet. This is particularly important in a country like India, where food insecurity and dietary imbalances are prevalent in many regions (Rao *et al.*, 2020).

Strategies and Approaches in Rice Biofortification

Biofortification, the process of enhancing the nutritional content of crops through breeding or biotechnology, presents a promising solution to address this nutritional deficiency. In this paper, we explore various strategies and approaches employed in rice biofortification to improve the nutritional quality of this vital crop.

1. Conventional Breeding Approaches

a. *Marker-Assisted Selection (MAS):* MAS involves the use of molecular markers linked to desirable traits, such as high micronutrient content, to expedite the breeding process. This approach enables breeders to select plants with the desired traits more efficiently, accelerating the development of biofortified rice varieties (Senguttuvel *et al.*, 2023).
b. *Hybridization*: Cross-breeding different rice varieties with naturally high nutrient content can lead to the development of new varieties with improved nutritional profiles. Hybridization programs focus on combining desirable traits from different parental lines to produce offspring with enhanced nutritional value (Qamar *et al.*, 2019).

2. Genetic Engineering Techniques

a. *Transgenic Approaches*: Genetic engineering allows scientists to introduce specific genes responsible for synthesizing or accumulating essential micronutrients into rice varieties. For instance, genes encoding enzymes involved in the biosynthesis of micronutrients like iron, zinc, and vitamin A can be inserted into rice genomes to enhance their nutritional content (Masuda *et al*., 2008).

b. *Genome Editing*: Technologies such as CRISPR-Cas9 enable precise modification of the rice genome, allowing targeted alterations to genes involved in nutrient metabolism. Genome editing offers a powerful tool for enhancing the nutritional quality of rice while minimizing unintended genetic changes (Senguttuvel *et al*., 2023).

3. Agronomic Practices

a. *Soil Nutrient Management*: Optimizing soil nutrient levels through fertilization and soil amendment strategies can influence the micronutrient content of rice grains. Balancing soil pH, providing micronutrient-rich fertilizers, and employing sustainable farming practices can enhance nutrient uptake by rice plants and improve grain nutritional quality.

b. *Water Management*: Proper water management practices, such as controlled irrigation and water-saving techniques, can influence nutrient availability and uptake by rice plants. Efficient water use ensures optimal nutrient absorption, contributing to improved grain quality.

4. Post-Harvest Interventions

a. *Food Processing Techniques*: Processing methods such as parboiling, milling, and fortification can help retain or enhance the nutritional content of rice grains. Fortification involves adding micronutrients to rice during processing to compensate for any nutrient losses and further boost its nutritional value.

b. *Bioavailability Enhancement*: Enhancing the bioavailability of micronutrients in rice through cooking methods or food combinations can improve their absorption in the human body. Techniques like soaking, germination, and fermentation can increase the availability of essential nutrients, promoting better nutritional outcomes (Samtiya *et al*., 2021).

In conclusion, biofortification in rice holds immense promise for addressing vital nutritional deficiencies and downgrading malnutrition on a global scale. Biofortification of rice represents a transformative approach to enhance nutritional security and combat malnutrition. By employing a combination of breeding techniques, genetic engineering, agronomic practices, and post-harvest interventions, researchers and agricultural stakeholders can develop biofortified rice varieties with enhanced nutritional contents. By leveraging advances in agricultural science and technology, biofortified rice varieties have the potential to improve the nutritional quality of diets, enhance health outcomes, and contribute to sustainable development. By increasing the availability of essential nutrients in rice, biofortification can improve the health and well-being of millions of people, particularly in regions where rice is a dietary staple. Continued investment in research, development, and promotion of biofortified rice varieties is essential to realizing the full potential of this innovative approach to addressing nutritional deficiencies and promoting food security. However, realizing this potential will require concerted efforts from governments, researchers, policymakers, and other stakeholders to overcome technical, regulatory, and socio-economic challenges and ensure equitable access to biofortified rice for those most in need. Through collaborative action and a commitment to innovation and inclusivity, biofortification can play a transformative role in building a healthier, more resilient, and food-secure future for all.

References

Bouis, H. E. 2003. Micronutrient fortification of plants through plant breeding: can it improve nutrition in man at low cost? Proceedings of the Nutrition Society, 62(2), 403-411.

Bouis, H. E., and Saltzman, A. 2017. Improving nutrition through biofortification: a review of evidence from HarvestPlus, 2003 through 2016. *Global food security,* **12**: 49-58.

Das, K., Roy, P., & Tiwari, R. K. S. 2023. Biofortification of Rice, An Impactful Strategy for Nutritional Security: Current Perspectives and Future Prospect. DOI: 10.5772/intechopen.110460

Dipti, S. S., Bergman, C., Indrasari, S. D., Herath, T., Hall, R., Lee, H.and Fitzgerald, M. 2012. The potential of rice to offer solutions for malnutrition and chronic diseases. *Rice*, **5,** 1-18.

Garg, M., Sharma, N., Sharma, S., Kapoor, P., Kumar, A., Chunduri, V. and Arora, P. 2018. Biofortified crops generated by breeding, agronomy, and transgenic approaches are improving lives of millions of people around the world. Frontiers in Nutrition, 12. https://doi.org/10.3389/fnut.2018.00012

Gregorio, G. B. 2002. Progress in breeding for trace minerals in staple crops. *Journal of Nutrition*, **132**(3), 500S-502S.

Majumder, S., Datta, K. and Datta, S.K. 2019. Rice biofortification: high iron, zinc, and vitamin-A to fight against "hidden hunger". *Agronomy*, **9**(12), 803.

Masuda, H., Suzuki, M., Morikawa, K. C., Kobayashi, T., Nakanishi, H., Takahashi, M., and Nishizawa, N. K. 2008. Increase in iron and zinc concentrations in rice grains via the introduction of barley genes involved in phytosiderophore synthesis. *Rice*, **1**(1), 100-108.

Müller, O. and Krawinkel, M. 2005. Malnutrition and health in developing countries. *Cmaj*, **173**(3), 279-286.

Ofori, K. F., Antoniello, S., English, M. M. and Aryee, A. N. 2022. Improving nutrition through biofortification–A systematic review. *Frontiers in Nutrition*, **9**, 1043655.

Ortiz-Monasterio, I., & Graham, R. D. 2000. Breeding for trace minerals in wheat. *Food and Nutrition Bulletin*, **21**(4), 392-396.

Paine, J. A., Shipton, C. A., Chaggar, S., Howells, R. M., Kennedy, M. J., Vernon, G. and Drake, R. 2005. Improving the nutritional value of Golden Rice through increased pro-vitamin A content. *Nature biotechnology*, **23**(4), 482-487.

Peña-Rosas, J. P., Mithra, P., Unnikrishnan, B., Kumar, N., De-Regil, L. M., Nair, N. S. . and Solon, J. A. 2019. Fortification of rice with vitamins and minerals for addressing micronutrient malnutrition. *Cochrane Database of Systematic Reviews*, (10).

Qamar, Z. U., Hameed, A., Ashraf, M., Rizwan, M. and Akhtar, M. 2019. Development and molecular characterization of low phytate basmati rice through induced mutagenesis, hybridization, backcross, and marker assisted breeding. *Frontiers in Plant Science*, **10**, 1525.

Samtiya, M., Aluko, R. E., Puniya, A. K. and Dhewa, T. 2021. Enhancing micronutrients bioavailability through fermentation of plant-based foods: A concise review. *Fermentation,* **7**(2), 63.

Sandhu, R., Chaudhary, N., Shams, R., Singh, K. and Pandey, V. K. 2023. A critical review on integrating bio fortification in crops for sustainable agricultural development and nutritional security. *Journal of Agriculture and Food Research*, **14,** 100830.

Sanjeeva Rao, D., Neeraja, C. N., Madhu Babu, P., Nirmala, B., Suman, K., Rao, L. S. and Voleti, S. R. 2020. Zinc biofortified rice varieties: challenges, possibilities, and progress in India. *Frontiers in Nutrition*, **7**, 26.

Saunders, J. and Smith, T. 2010. Malnutrition: causes and consequences. *Clinical Medicine*, **10**(6), 624.

Senguttuvel, P., CN, N., SV, S. P., LV, S. R., AS, H., RM, S. and Govindaraj, M. 2023. Rice biofortification: breeding and genomic approaches for genetic enhancement of grain zinc and iron contents. *Frontiers in Plant Science*, **14**, 1138408.

Sood S, Chaudhary DR, Jhorar P and Rana RS 2023. Biofortification: an approach to eradicate micronutrient deficiency. *Front. Nutr*. **10**:1233070. doi: 10.3389/fnut.2023.1233070

Talsma, E. F. and Pachón, H. 2017. Biofortification of crops with minerals and vitamins. World Health Organization, Rome. https://www. who. int/elena/titles/bbc/biofortification/en.

Trijatmiko, K. R., Dueñas, C., Tsakirpaloglou, N., Torrizo, L., Arines, F. M., Adeva, C., and Slamet-Loedin, I. H. 2016. Biofortified indica rice attains iron and zinc nutrition dietary targets in the field. *Scientific Reports*, **6**(1), 19792.

Venkatesh, U., Sharma, A., Ananthan, V. A., Subbiah, P. and Durga, R. 2021. Micronutrient's deficiency in India: a systematic review and meta-analysis. *Journal of Nutritional Science*, **10**, e110.

Wells, J. C., Sawaya, A. L., Wibaek, R., Mwangome, M., Poullas, M. S., Yajnik, C.S. and Demaio, A. (2020). The double burden of malnutrition: aetiological pathways and consequences for health. *The Lancet*, **395**(10217), 75-88.

World Health Organization and Food and Agriculture Organization of the United Nations. 2018. Guidelines on food fortification with micronutrients. World Health Organization.

Yadava, D. K., Hossain, F. and Mohapatra, T. 2018. Nutritional security through crop biofortification in India: Status & future prospects. *The Indian Journal of Medical Research,* **148**(5), 621.

2

Physiological, Biochemical and Molecular Basis of High Grain Protein Content

Introduction: N metabolism

Nitrogen is one of the three essential macronutrients for plant growth and yield (Su *et al.,* 2005) and most crucial elements for root growth also (Fageria, 2014). Most macromolecules and a variety of secondary and signalling substances, such as proteins, nucleic acids, cell wall components, hormones, and vitamins, contain nitrogen (N), an important element. In order to assimilate nitrogen, nitrate must be reduced to ammonium, and then ammonium must be converted into amino acids. Perez *et al.* (1973) pointed out that the rice genotype with high protein yield translocated N from leaf blade to the developing grains more efficiently than the low protein genotype. This was reported due to a high concentration of free amino acids in the sap translocated to the developing grains rather than to differences in translocation rates. Therefore, proper understanding of N metabolism and amino acid translocation are prerequisite to comprehend higher grain protein in high yielding rice genotypes. Use of N by plants involves several steps including uptake, assimilation, translocation and remobilization; and recycling and remobilization when the plant is ageing (Fig. 1). The enzyme nitrate reductase (NR) assimilates nitrate, the main form of nitrogen available to agricultural plants in the field. This enzyme and the nitrogen status of various higher plant systems have a strong positive link, and growth, yield, or protein content can occasionally be connected to this enzyme's level in seeds or leaves. Nitrate intake occurs at the root level, and it has been demonstrated that plants have two nitrate transport systems that work together to cohabit, absorb, and distribute nitrate from the soil solution throughout the entire plant. (Srivastava 1980) Following NO^{3-} uptake by NO^{3-} transporters and reduction of $NO3^-$ to $NO2^-$ by nitrate reductase (NR) in the cytoplasm, the process of $NO3^-$ assimilation commences (Lea and Miflin, 1974). The nitrite reductase then converts $NO2^-$ to NH4+ in plastids (NIR) (Xuan *et al.,* 2017).

Nitrogen is also taken up in the form of ammonium ions (NH_4^+) in rice since it is majorly grown under water logged condition and its metabolism requires less energy than that of nitrate (NO_3^-) (Bloom *et al.,* 1992). Ammonium transporters present across the membranes of root hairs (elongation of epidermal cells) are responsible to uptake the NH_4^+ from soil solution and import it into the plant system. There are different types of transporters involved in this uptake. Some of them have high while others have lower affinity to NH_4^+. After the $NH_4^+$is uptaken by the transporter into the epidermal cytosol, it combines with glutamate in the presence of glutamine synthetase (GS) to form one molecule of glutamine carrying two N atoms. Later this glutamine combines with 2-oxoglutamate in the presence of GOGAT (Glutamine: Oxoglutarate Amino Transferase) to form two molecules of glutamate, one of which serves as substrate for GS while the other is available for transport, storage or further metabolism. The reaction is catalyzed by NADH- GOGAT in cytosol and Fd GOGAT in plastids. The 2- Oxo glutamate used in the reaction is provided from mitochondrial NAD- isocitrate dehydrogenase. This completes the GS/ GOGAT pathway for assimilation of N in rice. The NADH GOGAT is strongly induced by glutamine being produced from the reaction catalyzed by GS. The GS2 assimilates any NH_4^+ produced through photorespiration in leaf and is immediately converted to glutamate through GS/ GOGAT cycle in leaf as per the following equations.

Glutamate + NH_4^+ $\rightarrow$ Glutamine +H_2O + ADP +Pi (Catalyzed by GS)

Glutamine + 2- oxoglutarate $\rightarrow$ 2 Glutamate + NAD+ (Catalyzed by NADH/ Fd GOGAGT)

Nitrogen is further transported to the shoot tissues through xylem vessels either in the form of glutamine or asparagine after conversion of glutamate from GS /GOGAT cycle in the presence of amino transferase. After uptake,NH_4^+ takes almost 1 hour to translocate to all other parts of the plant.Younger leaves are stronger N sink than the older leaves. After reaching the shoot parts, N that was transported in form of glutamine or asparagines is either utilized for further metabolism, stored as storage protein, structural proteins or synthesis of enzymes (Fig.2).

Nitrate Reductase (NR): Nitrate reductase (NR EC 1.7.1.1-3) is a molybdoenzyme that is involved in the synthesis of nitric oxide in plants. NR contains molybdenum as a cofactor (Moco) and is classified as an enzyme that participates in two-electron transfer reactions in all carbon, nitrogen, and sulphur cycles (Hille,1996). The assimilation of N-NO3 by plants requires the enzyme

NR, which is crucial (Epstein and Bloom, 2005). Three steps are involved in the reduction of nitrate to nitrite by NR: the first is a reductive half-reaction in which electrons are transferred from nicotinamide adenine dinucleotide phosphate (NAD(P)H) to reduce FAD, the second is electron transfer through the intermediate cytochrome b5 domain, and the third is an oxidative half-reaction in which electrons are transferred from the Mo centre to nitrate which converts to Nitrite (Skipper *et al.,* 2001). Environment where ammonium predominates, rice plants find it challenging to synthesise NR enzyme in its initial growth phase. Rice with high percentage of protein generally shows higher tendency to translocate more leaf N to the developing grains than the rice varieties with average grain protein content. Rice varieties with varying levels of grain protein did not differ in the level of N fractions and of enzyme activities in 10-days old seedlings. Nitrite reductase activities were present primarily in the shoot and were higher in the O^{3-} grown seedling (Marwaha *et al.,* 1976). According to Galangau *et al.* (1988) and Vincentz, Moureaux, and Leydecker (1993) NR enzyme activity in leaves and roots can be raised by the presence of higher NO^{3-}, although this higher activity did not guaranteed higher grain yield. Celestino (2006) also found that the NR enzyme activity in the leaves and roots could be stimulated by the presence of the substrate (NO^{3-}) and noticed a decrease in the NR activity in rice crop, grown in the absence of nitrogen fertilization. The peak of NR activity attained when NO^{3-} was most readily available. Pacheco *et al.* (2011) also obtained similar outcomes. NR activity in the leaves was shown to have increased by 2.09 folds when treated with NH^{4+}: NO^{3-} as opposed to NH4 in three rice cultivars Nanguang, Yunjing 38, and 4007 (Duan *et al.,* 2006). Ali *et al.* (2007) observed that NR activity fell significantly in the absence of light, it dramatically rose when nitrate and light were present. Santos *et al.* (2007) studied the ability of the "Piaui" variety to absorb and store NO^{3-} as well as its decreased NR activity in initial growth-stage may be the mechanisms for this variety to produce a large amount of N in grains. Li *et al.* (2008) examined while there was no discernible difference between the cultivars in terms of root NRA, Yangdao 6's leaf NRA was much greater than Nongken 57's. For both cultivars, the nitrification processes were most active in the rhizosphere soil, then in the bulk soil and on the root surface. Higher gene expression and enzyme activity related to N metabolism of rice at the booting stage affected secondary assimilation of N metabolism, which in turn led to greater grain development (Yi *et al.,* 2008). Kaur *et al.* (2015) observed protein and amino acid levels increased in all genotypes with an increase in nitrogen rate, which was accompanied by a notable rise in nitrite reductase activity. The activity pattern of the investigated enzymes showed an upward tendency from tillering to anthesis stage and then a downward trend after that, declining along with the reduction in protein and amino acid levels. Nitrate reductase activity significantly increased with an increase in nitrogen rate, which led to a rise in the levels of protein and amino acids in all genotypes observed in an experiment conducted by Kaur *et al.* (2016). A positive association of nitrogen assimilatory enzymes (nitrate reductase) with NUE and nitrogen content was also found, indicating that these may be the rate limiting enzymes in metabolism. Activity pattern of these enzymes revealed an increasing trend from tillering to anthesis stage and then declined that is parallel to the decreasing pattern in protein and amino acid contents. The activity of the NR enzyme was higher in plants grown in low acidic soils and fertilized with calcium nitrate. Higher enzyme activity related to rice N metabolism during the booting stage affected secondary assimilation of N metabolism, leading to greater grain development (Yi *et al.,* 2019).

Nitrite reductase (NiR): Nitrite, produced in cytosol, is transported into chloroplasts in leaves, where it is further reduced to ammonium ions by nitrite reductase (Lea *et al.,* 1993). Ten-day-old seedlings of three rice varieties with varying levels of grain protein did not differ in the level of N fractions and of enzyme activities, which was consistent with their variations in grain protein content. Nitrite reductase activities were present primarily in the shoot and were higher in the O3-grown seedling (Marwaha *et al.,* 1976). Ali *et al.* (2007) reported NiR activity increased threefold when nitrate and light were present, there was no influence on NiR activity when light was absent. Nitrite reductase activity significantly increased with an increase in nitrogen rate, which led to elevation of protein and amino acid content across all genotypes. Activity of this enzyme showed an upward tendency from tillering to anthesis, then a downward trend that dropped with the higher levels of protein and amino acids (Kaur *et al.,* 2015)

Glutamine synthetase (GS): The glutamine synthetase (GS) enzyme catalyses the synthesis of glutamine from glutamic acid and newly absorbed ammonium, as demonstrated by Arima and Kumazawa (1975) using the ir 5Ni.trace r technique (EC 6.3.1.2). Glutamine is synthesised by the glutamine synthetase (GS)/glutamate synthetase (GOGAT) cycle (Lea and Miflin, 2003). During the first week of germination, the activity of glutamine synthetase in root tended to be lower than that of the NH4 growing seedling. Marwaha *et al.* (1976) did not find any difference in N fraction levels or enzyme activities in three rice varieties with varying levels of grain protein in ten-day-old seedlings. Two types of GS, cytosolic GS1 and chloroplast GS2, were characterized by biochemical studies (Lea *et al.,* 1990). GS1 is presumably involved in the generation of glutamine for primary nitrogen assimilation in roots and for intercellular nitrogen transport in vascular bundles, whereas GS2 functions in the reassimilation of photorespiratory ammonia in leaves. GS2 gene was reported to play role in N-translocation and mainly active at flag leaf at flowering stage. Two glutamine synthetases (GS1 and GS2) were identified in green rice leaves by chromatography on a DEAE-cellulose column. Contribution of GS2 to the total activity was high compared to GS1. After 48 hours of exposure to light, the usual pattern of green leaves was established. GS1 dropped as GS increased during the process of etiolated leaves turning green. Re-chromatography of GS1 and GS2 didi not show any interconversion (Guiz *et al.,* 1979). Simpson *et al.* (1981) investigated throughout the grain-filling cycle and observed that glutamine synthetase activity in the flag leaf decreased but the enzyme activity reached a peak in the flag leaf's senescence process. According to observations noted by Kamachi *et al.* (1991) the GS1 in senescent rice leaf blades produces glutamine, which was then transported to the developing tissues. During the 35 days of senescence, total GS activity decreased to less than a fifth of its starting level, with a drop in the amount of GS2 polypeptide being the primary contributor. GSI is crucial for the export of leaf nitrogen from senescing leaves. Yamaya *et al.* (1992) examined GSI protein concentration by on a unit of g fresh weight basis, the indica cultivars Chinsurah Boro I and Blue Stick outperformed the japonica cultivar Sasanishiki in the senescing leaf blades.

Glutamine synthetase cycle is crucial for the build-up of anaerobic amino acids. The accumulation of amino acids in the roots of 3-day-old rice seedlings that had been exposed to anaerobic conditions for 48 hours was examined. After 8 hours of anaerobic treatment, these activities increased, and it was demonstrated using immunoprecipitation of 35S-labeled proteins that the therapy resulted in the synthesis of glutamine synthetase and ferredoxin-dependent glutamate synthase (Reggiani *et al.,* 2000). Several studies emphasized the role of GS1 in N management, growth rate, yield and grain filling (Martin *et al.,* 2006). Higher GS activity in both its roots and leaves at 42 DAG were displayed in 'Piaui' variety. Only when this variety was grown with 200 mg NO-N L1 did the GS activity in the shoot increase. In both treatments (20 and 200 mg NO3-N L-1), the roots of the "Piaui" variety had stronger GS activity than the "IAC-47" variety. In both kinds, the GS activity in the roots was twice as high as that in the shoots, demonstrating that roots are where plants prefer to receive and decrease NO^3 (Santos *et al.,* 2007) (Fig. 2). Under high temperature condition, at the grain-filling stage, GS activity drastically decreased. It is hypothesised that GS may not be the primary enzyme in regulating glutamine concentration in grains, as the decline in GS activity in grains had no effect on the deposition of amino acids and protein (Cheng-gang *et al.,* 2011). Nitrogen assimilatory enzymes and glutamine synthetase had a positive association with NUE and nitrogen content, observed by Kaur *et al.* (2016) suggesting that these enzymes may be the rate-limiting ones in metabolism.

Glutamate synthase (GOGAT): Glutamate synthase (GOGAT) catalyses the transfer of the amide group of glutamine produced by glutamine synthetase (GS; EC 6.3.1.2) to 2-oxoglutarate to produce two molecules of glutamate. According to Lea and Mi, (1974) the GS/ GOGAT cycle is the primary mechanism for NH4+ assimilation in plants under normal metabolic conditions. One of the glutamate molecules can be employed in a variety of synthetic processes, while the other can be recycled as a substrate for the GS reaction (Lea *et al.,* 1990; Sechley *et al.,* 1992). Glutamate-synthase activities declined during grain-filling period measured in the flag leaf and attained a peak very late in the course of senescence of the flag leaf (Simpson *et al.,* 1981).

NADH-GOGAT is responsible for the production of glutamate from glutamine which travels through the circulatory system. To investigate the roles of NADH-GOGAT in the nitrogen remobilization and reutilization in rice, several cultivars of indica and javanica rice were studied to determine the amounts of NADH-GOGAT proteins in senescing and developing leaf blades, respectively. On the other hand, NADH-GOGAT content in the young leaf blades of Sasanishiki was the highest (Yamaya *et al.,* 1992). They observed increased activity and protein concentrations of NADH GOGAT in the 10th leaf blade before emergence. When compared to the 10th leaf, the unexpanded, nongreen part of the 9th leaf blade had more than half the amount of NADH-GOGAT protein and activity. On a fresh weight basis, the growing, green area of the 9th leaf blade outside of the sheath contained a somewhat lower abundance of NADH-GOGAT protein than the area that was not green. Age-related declines in NADH-GOGAT levels were observed in fully expanded leaf blades below the 9th leaf, with the oldest 5th leaf blade having 4% fewer of the protein than the youngest 10th leaf blade. Hayakawa *et al.* (1993) reported during the ripening process, nitrogen accumulation in the apical spikelets on the primary branches of the main stem of rice plants has been investigated (0-35 d after flowering). In the first 15 days following flowering, the amount of GOGAT protein and its activity increased 4- and 6-fold, respectively. At that time, when the spikelets were just starting to gain dry weight and assemble storage proteins, the highest levels of NADH-GOGAT were discovered. These findings indicate that in rice plants, NADH-GOGAT is in charge of synthesising glutamate from the glutamine that is delivered to the spikelets from senescent tissues. Reggiani *et al.* (2000) reported glutamate synthase cycle play an important role in anaerobic amino acid accumulation. Cheng-gang *et al.* (2011) studied the effect of high temperature on dynamic changes of glutamate synthase (GOGAT) activity in grains and observed higher soluble protein content in grains. It was suggested that GOGAT in grains played important roles in nitrogen metabolism at the grain-filling stage and responsible for higher protein content in grains.

In the model monocotyledonous plant rice (*Oryza sativa* L.), FADH-GOGAT represents the major activity of GOGAT and has been biochemically characterized (Suzuki and Gadal, 1982). Major functions in photorespiration and primary nitrogen assimilation are played by ferredoxin-dependent glutamate synthase (FADH-GOGAT). According to Zeng *et al.* (2017), OsFd-GOGAT may help with nitrogen remobilization during leaf senescence, which may increase rice's ability to use nitrogen effectively.

Remobilization of N in Grains

During the grain filling stage, N that was stored in the leaves is remobilized to the developing seed, which poses a strong sink for photosynthates and nutrients. However, a negative association was found between grain protein concentrations and yield (Beninati and Busch, 1992). Remobilizing efficiency, however is highly genotype dependent (Kichey *et al.,* 2007) indicating the scope of manipulating N remobilisation to improve the nutritional quality of grain without reducing the yield potentiality. However, N remobilization is not prompted by this stage and is a continuous process that is triggered by each emerging organ that provides a potential sink for N viz. N fluxes from older leaves to younger leaves during vegetative phase. Remobilization is always associated with higher protease activity and leaf senescence.

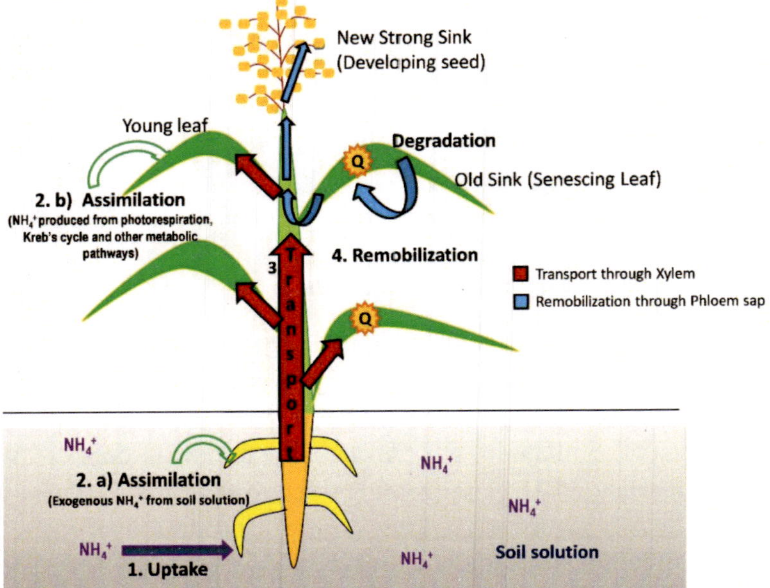

Fig. 1: Represents overall mechanisms involved in N- metabolism in rice plant. Q represents Glutamine.

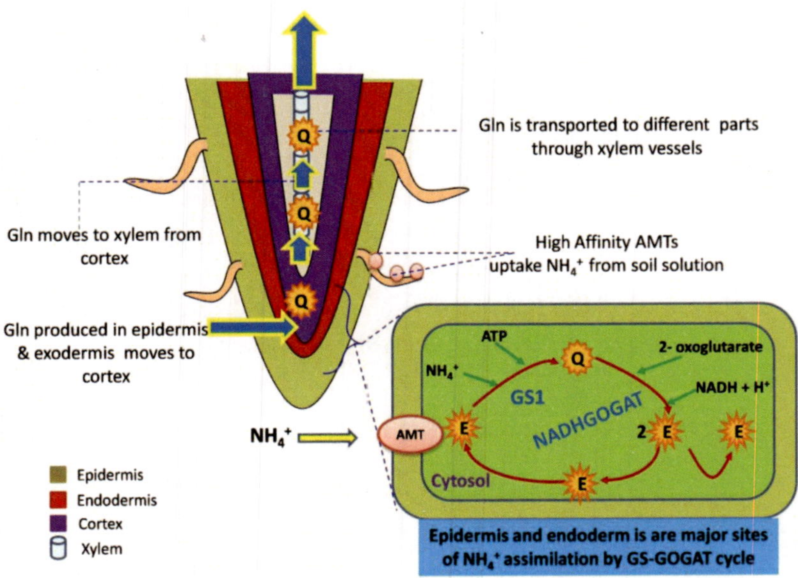

Fig. 2: Assimilation of N in rice root.

(Q, Glutamine; E, Glutamate; AMT, Ammonium transporter; GS, Gutaminesynthetase; GOGAT Glutamine:oxoglutarate amino transferase)

The remobilization efficiency of N mainly depends on sink strength of seed, efficiency of transfer processes located in the source such as leaves, stems and panicles as well as on phloem pathway efficiency and also growth environment of the plant. It preferentially occurs in the form of glutamine through phloem sap which accelerates with leaf aging. Tilsner *et al.*, (2005) observed that the concentration of free amino acids in the phloem sap was in the range of 50–200 mM during grain filling and apoplasmic phloem loading depended on amino acid transporters in the sieve elements and companion cell complexes. It is an energy dependent process. Membrane bound amino acid transporters carry amino acids into sieve element / companion cell complexes of minor veins. Amino acid permeases (AAP), being non-specific to amino acids, are engaged in the phloem loading process (Fischer *et al.,* 2002). The AAP are localised on plasma and internal membranes of main viens of mature leaves. Specific amino acid transporters are also reported by Hirner *et al.,* (2006) and van der Graaff *et al.* (2006) which may be involved in the phloem loading process.

Phloem loading is the key step in the process of N remobilization. Amino acid and peptide transporters which are expressed in the seed involve in endosperm or cotyledon loading of N compounds during embryogenesis (Rentsch *et al.,* 2007; Tsay *et al.,* 2007). Supplying of N in seed mostly depends on efficiency of phloem loading by amino acid transporters in source organs and also on sink strength of developing seed.

Asparagine is one of the major form of N in the phloem sap, critical for N assimilation, distribution and remobilization at the whole plant level (Lea *et al.,* 2007). Asparagine synthetase (ASN) transfers an amide group from glutamine to aspartate and forms asparagine, functions in N assimilation, remobilization and in glutamate and glutamine recovery (Gaufichon *et al.,* 2013, Moison *et al.,* 2018). Gaufichon *et al.* (2013) observed that disruption of ASN2 results in asparagine depletion in phloem saps, lower 15N flux from source leaves to sinks, delayed senescence and abnormal growth phenotypes, suggesting the significant role of ASN2 in N assimilation, distribution and remobilization.

Amino acids to be stocked in phloem for N-remobilization are produced by proteolysis of proteins synthesized before the onset of reproductive phase (Patrick and Offler, 2001). Rubisco being the major soluble protein provides the major source of N for remobilization along with other photosynthesis related proteins. Proteases for the proteolysis of proteins are present in cytosol and cell sap. Over production of reactive oxygen species from chloroplasts (major), peroxisomes or mitochondria, (Zimmermann and Zentgraf, 2005) during aging triggers the process of proteolysis. Several other proteolytic mechanisms operating in different cell compartments may be involved in N-remobilization. Bassham *et al.,* 2006 reported the role of autophagic recycling for N- mobilization during grain filling.

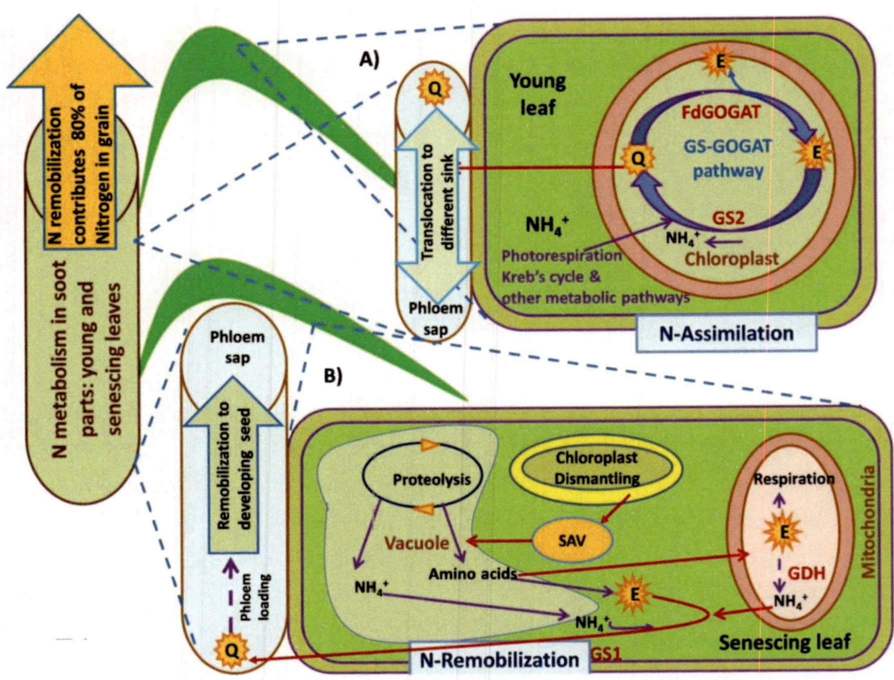

Fig. 3: (A) Represents mechanism of N- assimilation and transport in Young leaves; **(B)** Represents mechanism of Degradation, N-remobilization and transport to developing seed from senescing leaves. SAV stands for Senescence-associated vacuole, Q represents Glutamine, E represents Glutamate, GS represents Glutaminesynthetase, GOGAT for Glutamine : oxoglutarate amino transferase, GDH Glutamate dehydrogenase. Adapted from Masclaux-Daubresse *et al.*, 2010.

Glutamate dehydrogenase (GDH) may also have some role in N assimilation (Lea and Miflin, 2011), but it seems to predominantly function in glutamate deamination and ammonium supply for GS1 (Moison *et al.,* 2018) during N remobilization. GDH seems to predominantly function in glutamate deamination and ammonium supply for GS.

Through the glutamine synthetase/glutamine:2-oxoglutarate amidotransferase (GS/GOGAT) cycle, plants convert inorganic nitrogen received from soil into organic forms like Gln and Glu. A crucial enzyme involved in the nitrogen metabolism of higher plants is glutamine synthetase (GS). GOGAT facilitates the transfer of an amide group from Gln to 2-oxoglutarate to make two molecules of Glu, as opposed to GS, which catalyses the production of Gln from Glu and ammonia. However, it is not fully known how the GS/GOGAT cycle regulates the carbon-nitrogen balance.

Several studies emphasized the role of GS1 in N management, growth rate, grain filling and yield (Hirel *et al.,* 2001; Obara *et al.,* 2004; Tabuchi *et al.,* 2005; Martin *et al.,* 2006). However, the mobilization process still remains complex due to involvement of several isoforms of enzymes, unknown functions of proteins and multidimensional activity of enzymes. In rice, three GS1-encoding genes (OsGS1;1, OsGS1; 2, OsGS1;3) was identified (Ishiyama *et al.,* 2004) with its important roles in growth rate and grain filling (Figure 3). OsGS1;1 gene product, which is located in companion cells and parenchyma cells of leaf tissues, is responsible for the generation of glutamine for remobilization through the phloem. Both germination and senescence require huge amount of protein remobilisation and in both the processes show higher activity of amiotransferases. Role of transaminases have also been documented during senescenece leading to greater accumulation of amino acids like tyrosine, leucine, isoleucine anthe non-proteic amino acid, c-aminobutyric acid (GABA) with aging thereby reducing the concentration of glutamate which was abundant in young leaves (Diaz *et al.,* 2008).

Transaminases are reversible pyridoxal-5¢-phosphate containing enzymes that catalyse the transfer of ana-amino group to 2-oxo acids. Forde and Lea (2007) has postulated that the reversibility of their enzymatic activity accounts for the relative stability of glutamate concentrations in plant tissues.

Molecular Aspect of N and Amino Acid Assimilation

Nitrogen (N) is important for the growth and production of plants. However, the use of nitrogenous fertilizers causes issues with the economy and the ecology. The creation of rice cultivars with high nitrogen use efficiency (NUE) is essential for decreasing environmental problems and attaining sustainable agriculture in order to meet the rising global food demand.

In rice (*Oryza sativa* L.) leaves, glutamine synthetase (GS) and its isoforms have been widely researched, but little is known about GS in the spikelets and rachis of rice panicles. Major functions in photorespiration and primary nitrogen assimilation are played by ferredoxin-dependent glutamate synthase (Fd-GOGAT, EC 1.4.7.1). However, because no mutant or knockdown lines of OsFd-GOGAT in rice (*Oryza sativa* L.) have been described, the knowledge of OsFd-role GOGAT's in rice foliar nitrogen metabolism is still limited. GS activity and its isoforms were determined in the spikelets, rachis, and flag leaves during grain development. The GS activities per unit fresh weight of these tissues reached maximum levels around the milk stage of the endosperm and declined thereafter as grain filling progressed. Flag leaves had 5-fold higher GS activity per unit fresh weight than rachis and 7-fold higher than spikelets at the milk stage. Native-PAGE and activity staining indicated that there were two GS isoforms in the rachis and they were the same as the cytosolic form (GS 1) and the chloroplastic form (GS2) in leaves, respectively. The GS2 was the dominant isoform in flag leaves, whereas GS 1 was the dominant isoform in the rachis. However, only GS 1 activity was detectable in spikelets including hull and kernel. The lack of GS2 activity in the spikelets was consistent throughout the entire grain development stage and was confirmed by protein immunoblotting (Zhang *et al.,* 2000).

The high affinity NH_4^+ transporter is encoded by Ammonium Transporter 3 Member 1 (AMT1-3), which is primarily expressed in rice roots during nitrogen deficiency and is known as a high-affinity transport system (HAT). Ammonium (NH_4^+) is the main source of nitrogen for rice (*Oryza sativa* L.). OsAMT1;1, a key member of the OsAMT1 gene family involved in NH_4^+ transport in rice plants, controls the majority of the NH_4^+ uptake by roots. The NRT gene families' members play a major role in mediating the transfer of NO^{3-}. The high-affinity NO^{3-} transporters belong to the NRT2 family (Islam 2019). NRT 2.2 is responsible for uptake and transport of N to shoot. *NRT2.1 is* responsible for NO^{3-} uptake from soil (Islam, 2019).

The expression patterns of three members of the AMT1 family of genes in rice seedling roots in response to altered nitrogen provision and diurnal changes in irradiance were examined by Kumar *et al.* (2003) in order to investigate the molecular basis of high-affinity ammonium absorption by rice plant roots (*Oryza sativa* subspecies *indica*). When plants acclimated to 10 mM external NH_4^+ for 3 weeks were switched to 10 mM NH_4^+, the NH_4^+ influx and transcript levels of OsAMT1.1 in roots fell several fold within 48 hours. Likewise, OsAMT1.1 was rapidly upregulated and 13 NH_4^+ influx was seen in the roots of plants that had been acclimated to 10 mM NH_4^+. Following these treatments, OsAMT1.2 transcript abundance changed about 50% less than OsAMT1.1, and OsAMT1.3 expression changed even less. In contrast, late in the photoperiod, OsAMT1.3 and NH_4^+ influx rose about threefold in response to daily fluctuations in irradiance, but OsAMT1.1 and OsAMT1.2 showed relatively minor changes. The findings implied that in response to variations in

N status and daily irradiance, the expression of three members of the OsAMT1 family of genes in the roots of rice seedlings was differently regulated to control high-affinity NH_4^+ inflow. The three OsGS1;1, OsGS1;2, and OsGS1;3 genes for cytosolic glutamine synthetase (GS1) are present in rice (*Oryza sativa* L.) plants. As opposed to OsGS1;2, and OsGS1;3, which were mostly expressed in roots and spikelets, respectively, OsGS1;1 was expressed in all organs tested with higher expression in leaf blades. According to the findings, GS1;1 is crucial for rice's proper growth and grain filling because GS1;2 and GS1;3 were unable to fill the GS1;1 function gap (Tabuchi *et al.,* 2005). The findings of Cai *et al.,* 2009 revealed increased total GS activities, soluble protein concentrations in leaves, higher total amino acid and total nitrogen content in the entire plant, indicating an elevated metabolic level in GS-overexpressed plants. Comparing seeds from GS-overexpressed plants to seeds from wild-type plants, it was found that GS-overexpressed plants produced less grain and total amino acids. In contrast to the other two types of GS-overexpressed plants, GS1;2-overexpressed plants showed higher sensitivity to salt, drought, and cold stress conditions and resistance to Blast selection, while the other two types of GS-overexpressed plants showed no discernible differences for these stress conditions when compared to wild-type plants.

In flag leaves during the active grain filling stage, gene expression analysis of ammonium transporter gene family members showed that the three OsAMT3 family members (OsAMT1;1-3), one OsAMT2 family member (OsAMT2;3), and the high affinity OsAMT1;1 were all differentially expressed and were influenced by different nitrogen doses. The OsAMT1;1, OsGS1;1, and OsGS1;2 expression levels in the flag leaves during the grain filling stage coincided with both increases and decreases in seed protein contents, indicating that high nitrogen nutrition promotes the repression of these genes, which may be crucial during grain filling (Gaur *et al.,* 2012). Bao *et al.* (2014). In rice roots, AMT1-3 encodes the high affinity NH4+ transporter, which is mostly expressed when nitrogen is scarce. Significant reductions in leaf carbon and nitrogen content were seen in AMT1-3-overexpressing plants, along with an increase in the leaf C/N ratio. In AMT1-3-overexpressing plants, significant alterations in soluble proteins and sugars were also noted. Genes involved in carbon and nitrogen metabolism have unique expression patterns, according to gene expression analyses. Furthermore, under both low and high nitrogen growth circumstances, AMT1-3-overexpressing plants showed a consistent link between the metabolites and gene expression patterns. We therefore postulated that the inadequate development and yield of transgenic plants was caused by the carbon and nitrogen metabolic imbalance generated by AMT1-3 overexpression (Bao *et al.,* 2015)

The identification of the ABC1 gene which produces a ferredoxin-dependent (Fd)- GOGAT in rice, in terms of its function. The accumulation of excessive amounts of amino acids with high N/C ratios (Gln and Asn) and several intermediates in the tricarboxylic acid cycle in abc1-1 as revealed by metabolomics research suggests that ABC1 is essential for nitrogen assimilation and the maintenance of the carbon-nitrogen balance. The ABC1 coding area had five non-synonymous single-nucleotide polymorphisms that were described as three different haplotypes and were highly and precisely differentiable between the japonica and indica subspecies. These findings by Yang *et al.,* (2016) imply that via regulating nitrogen uptake and the carbon-nitrogen balance, ABC1/ OsFd-GOGAT is crucial for plant growth and development. On chromosome 7, the *GOGAT1* locus was located in a 54.1 kb area. OsFd-GOGAT sequencing revealed one substitution (A to T) at the 3017th nucleotide of the open reading frame, resulting in the amino-acid change of leucine to histidine. The gogat1 mutant had a slower rate of seed germination while having considerably more grain protein than the wild type. The total amino acids in the top three leaves and the top internode increased significantly during the grain-filling stage. According to the study of Zeng *et al.,* 2017 findings, OsFd-GOGAT may help with nitrogen remobilization during leaf senescence, which may increase rice's ability to use nitrogen effectively. James *et al.* (2018) suggested that concurrent OsGS1;1 and OsGS2 isoform overexpression in rice improved physiological tolerance and agronomic performance under challenging abiotic stress conditions, seemingly working through numerous molecular pathways.

Storage Protein Accumulation in Endosperm

Rice caryopsis is developed from the fertilized pistil. Enveloped by the brownish pericarp, the dehulled grain is called brown rice. Next to the pericarp are two layers of cells named tegmen or seed coat. The embryo lies on the ventral side of the spikelet next to the lemma. The remaining part of the caryopsis is the endosperm which provides nourishment to the germinating embryo. The endosperm is wrapped by the aleurone layer beneath the testa (seed coat), having the starch storage parenchyma inside (Matsuo and Hoshikawa, 1993). Grain development is a continuous process and the grain undergoes distinct changes before it fully matures. After fertilization, endosperm cells proliferated and differentiated, lasting about 12 days. The dry matter such as starch and protein is accumulated strongly from three day after flowering

(DAF) to DAF12 (Rosario *et al.,* 1968; Xiong *et al.,* 2005). In contrary to the other crops, around 70-80% of the total storage protein in rice is glutelin (Yamagata *et al.,* 1982) and 20-30% is prolamin. Thus the composition and structure of glutelin and prolamin have the crucial role in ascertaining the nutritional traits in rice. Storage protein in the endosperm are mainly filled between the starch granules in the form of independent protein bodies(PBs), which are abundantly distributed in the surrounding cells and close to the cell wall (Fig.2), and the higher the rice protein content, the more PBs are packed and compacted around the starch granules. There are two types of PBs formed by RSPs, the spherical type I protein body (PB-I) with concentric-sheet structure and the ellipsoid type II protein body (PB-II) without sheet structure. PB-I accounts for about 20%–30% of the total RSPs, mainly prolamin (Kubota *et al*, 2010). Due to the strong hydrophobicity. that can form disulfide bonds among the polypeptides of prolamin, PB-Iparticles show a regular round and spherical morphology with a diameter of 1–2 μm and can hardly dissolve in water.Because of the insoluble and indigestible characteristics of prolamin, the PB-I does not change its properties after cooking, witha strong resistance to proteolytic enzymes, and is not absorbed and utilized by the human body.PB-IIcomprises about 65% of the total RSPs and is mainly composed of glutelin proteins and a small number of globulins, which are soluble proteins with weak physical properties and are readilydigested and absorbed (Tanaka *et al*, 1980). Glutelins are transported to the protein storage vacuole by way of the Golgi and form PB-II. PB-II is larger (3-4 μm), irregularly shaped, and of highly uniform staining density (Takemoto *et al.,* 2002).

Glutelin in rice is initially synthesized as a 57 kDa precursor and subsequently cleaved into two subunits, viz. a 37 to 39 kDa acidic and a 22 to 23 kDa basic subunits (Yamagata *et al.,* 1982). Glutelin genes can be classified into four subfamilies, GluA, B, C and D. The molecular masses of prolamins is ranging from 12 to 17 kDa and encoded by a complex multigene family (Mitsukawa *et al.,* 1999). Rice glutelin and prolamin genes are expressed generally during seed development. Investigation on the expression level in six members from the GluA and GluB subfamilies showed that the detectable expression level reaches at the peak activity at 10 days after flowering and gradually decline towards maturation (Duan and Sun, 2005). This pattern was found almost similar in other studies on GluA, GluB and GluC subfamily members (Okita *et al.,* 1989; Takaiwa and Oono, 1990, 1991; Mitsukawa *et al.,* 1998). Most of the seed-storage protein and starch synthetase genes are active during the grain development, indicating the commonality in regulation mechanism on the transcription level. Most of the starch synthetases also reach at the activity peak at 6 to 10 days after flowering. The only exception was the expression of the *Wx* gene which after declining statining from 6 days, rise again and peaked at 15 DAF before declining to trace level at 20 days after flowering (Duan and Sun, 2005). Although the expression pattern of *Wx* in rice grain is cultivar dependent (Wang *et al.,* 1995).

Identified QTLs for Grain Protein Content

More than hundred stable and consistent QTLs for GPC have been identified and mapped on all twelve chromosomes of rice (Fig. 4). Majority of them were mapped on chromosomes 1, 2, 6, 7, 10 and 11 (Table 1). In the beginning of this century, Tan *et al.,* (2001) first reported 2 QTLs for protein content in rice using a RIL population derived from Zhenshan 97/Minghui 63 cross. The *Wx* marker associated with amylose content (AC) was linked to one of the QTLs on chromosome 6. The QTL for protein content linked to ACs near to *Wx* (RM190) identified in another study explained 19.3% PVE (Yu *et al.,* 2009). Lou *et al.* (2009) identified two major QTLs for protein content on chromosomes 6 (RM588– RM540) and 7 (RM5436–RM6776). Employing double haploid population derived from an interspecific backcross (*O. sativa* × *O. glaberrima*) population, Aluko *et al.,* (2004) identified and mapped four QTLs explaining the phenotypic variance of 4.8 to 15.0%. Among those, one QTL, *pro6* was closely associated with *Wx* gene. In the last decade, a number of studies have been conducted to map the QTLs regulating GPC in rice. Zheng *et al.,* (2012) taking different grain filling stages mapped 10 unconditional QTLs and 6 conditional QTLs, explaining 8.53-19.59% and 8.76-23.70 % of PVE for GPC, respectively. Xu *et al.* (2015) detected one QTL *qPRO3.1* for protein content and another 29 QTLs for different grain quality traits. Using a chromosome segment substitution line (CSSL) population derived from the cross of a *Japonica* variety (Sasanishiki) with *Indica* variety (Habataki) Yang *et al.* (2015) identified seven QTLs, one of them was consistent (*qPC-1*) across three environments explaining 10.38-15.43% of PVE. F2 and F3 segregating populations from the cross between a SL402 (low protein CSSL) and Sasanishiki were used to delimit the region of *qPC-1* to a 41-kb on chromosome 1. These results could be helpful in introgression of the QTL for GPC into rice cultivars using marker assisted selection. Eight QTLs for PC included two QTLs on chromosome 6 (Kinoshita *et al.,* 2017) and a new QTL for PC and AC on chromosome 8 was reported.

In one study, Bruno *et al.* (2017) 34 using a DH population derived from a cross between Cheongcheong and Nagdong mapped a QTL for GPC on chromosome 7 and associated marker RM8261, explaining 14 % of PVE. As has been shown by previous studies, identification of robust QTLs for GPC in rice grains have been restricted because of lack of appropriate donors, non-utilization of high throughput phenotyping and genotyping platforms and high genotype × environment (G × E) interaction. To overcome these restrictions, recently Chattopadhyay *et al.* (2019) 35 genotyped a BC_3F_4 mapping population derived from the cross between grain protein donor, ARC10075 and high-yielding cultivar Naveen, using 40 K Affymetrix custom SNP array and identified one stable QTL *qGPC1.1* and another two stable QTLs for single grain protein content namely *qSGPC2.1* and *qSGPC7.1* explaining 13%, 14% and 7.8% of PVE, respectively. Based on genotyping-by-resequencing, a total of 14 QTLs controlling GPC were identified with an indica/japonica (HHZ/JZ1560) RIL population (Wu *et al.,* 2020). The qGPC10 was unique in this study. Seven of the fourteen QTLs were repeatedly identified across two years. The stably inherited *qGPC1-1* with large effect was validated and delimited to a ~862 kb region flanked by JD1006 and JD1075 on the short arm of chromosome 1 (Wu *et al.,* 2020).

QTLs identified in these study can be useful to improve the nutritional quality of rice grain. The closely linked markers that flanked the identified QTLs can be used to aid quality selection in breeding programs. And the results of the coincidence between the QTL detected, and the loci involved in protein biosynthesis pathways might be helpful for gene cloning by the candidate gene method.

In the past decade, although numerous QTLs for GPC variation have been detected in rice germplasms, few QTLs have been cloned except for *qPC1/OsAAP6* (Peng *et al.,* 2014). However, the *OsAPP6* expression level is found to be associated with GPC variation only in *indica* accessions, and no correlation between *OsAAP6* expression level and GPC variation was detected in the *japonica* genetic background. It seems that *OsAAP6* could be only used as a target gene to regulate GPC in *indica* breeding programs. Yang *et al.* (2019) identify two stable quantitative trait loci (QTLs), *qGPC-1* and *qGPC-10,* controlling GPC in a mapping population derived from indica and japonica cultivars crossing. Map-based cloning reveals that *OsGluA2,* encoding a glutelin type-A2 precursor, is the candidate gene underlying *qGPC-10*. It functions as a positive regulator of GPC and has a pleiotropic effect on rice grain quality. One SNP located in *OsGluA2* promoter region is associated with its transcript expression level and GPC diversity. Low expression type

(*OsGluA2^LET^*), mainly present in japonica accessions, originates from wild rice. However, high expression dominant type (*OsGluA2^HET^*) in indica, is acquired through mutation of *OsGluA2^LET^* (Yang *et al.,* 2019) suggesting that *OsGluA2^LET^ al*lele from japonica accessions could be directly replaced with *OsGluA2^HET^ al*lele to improve the nutrition quality through marker-assisted selection. Alam *et al.,* (2023) using populations derived from Kongyu 131/ Cypress (population-I) and Kongyu 131/Vary Tarva Osla (population-II) detected 25 QTLs for glutelin, prolamin, globulin and total protein content. A QTL, qGPC7.3 was delineated for total protein content and validated in introgression lines. Chattopadhyay *et al.,* (2023) through association mapping detected two unique QTLs for GPC-*qGPC 4.1* and *qGPC 4.2* associated with markers RM 17,600 and RM 1272, respectively were found co-localized with other two QTLs for amylose content, *qAC 4.1* and *qAC 4.2.* A set of 96 introgression lines (ILs) were developed (Yun *et al.,* 2016) from a cross between the Korean elite *O. sativa* japonica cultivar 'Hwaseong' and *O. rufipogon* (IRGC 105491). Five QTLs for protein content were identified in this study. Among those, qPC11 and qPC12-1 were detected in 83 ILs, while the other three QTLs, qPC2, qPC6, and qPC12-2 were found in 96 ILs. At all loci, *O. rufipogon* alleles increased the protein content.

QTLs for Amino Acid Content

In addition to GPC the improvement in the amino acid composition (AAC) determines qualitative enhancement of nutritional composition. Essential amino acids cannot be synthesized by animals, but play a crucial role in metabolism. Therefore, improving essential amino acid content in rice grain can substantially supplement nutrition of the people depends on rice as staple food. As compared to GPC, although number QTLs detected in essential amino acids are less, but those bi-parental and association mapping using various mapping populations have provided useful genetic information for improving the amino acid profiling in rice grains. Wang *et al.* (2007) reported the QTLs for AAC in milled rice, using RILs derived from a cross between Zhenshan 97 (*indica*) and Nanyangzhan (*japonica*). For all amino acids Zhenshan 97 was larger than that of Nanyangzhan in the 2-year experiments. In the study, several QTLs clusters for AAC were identified on chromosomes 1, 2, 3, 4, 7, 8, and 10. There were some large effect QTL for Asp, Thr, Gly, and Ala located in between RM472 and RM104 on chromosome 1 explaining phenotypic variance of 17.5–33.3% and QTL for Gly, Arg, Met, and Pro located in between RM125 and RM214 on chromosome 7 explaining PVE of 4.68– 7.69%. Probable functional genes inside the QTLs involved in amino acid metabolism pathways and protein biosynthesis. In another study, Zhong *et al.* (2011) reported 48 and 64 QTLs each contributing 4.0% to 43.7% PVE in subsequent years. A good coincidence was detected between the detected QTL and the loci involved in amino acid metabolism pathways, nitrogen assimilation and transport and protein biosynthesis. Zheng *et al.* (2008) mapped ten additive effect QTLs explaining 12-35 % of PVE for histidine on chromosomes 1, 2, 3, 6, 7, 10, 11, and 12 and 8 QTLs explaining 16–33% of PVE for arginine on chromosomes 2, 3, 5, 6, 7, 10, 11, and 12. Two QTLs for Histidine and 2 for Arginine content also showed significant dominant main effects from the triploid endosperm. Apart from that various interactions between QTLs and environment were identified for 5 QTLs associated with Histidine content and 2 QTLs associated with Arginine contents. Yoo (2017) identified QTL in brown rice for AAC including Ala, Val, Leu, Ile, and Phe which was the only QTL cluster on chromosome 3 using RIL population from Dasanbyeo (*indica*) and TR22183 (*japonica*) contributing 10.2-12.4 % PVE for the content of six amino acids. The QTL cluster (*qAla3, qVal3, qPhe3, qIle3,* and *qLeu3*) was found to be associated with the five amino acids content explaining PVE from 10.2 to 12.8%. In the same study, another QTL for Lys was identified on chromosome 3. QTL analyses using recombinant inbred lines from the cross between *indica* (Milyang 23 or M23) and *japonica* (Tong88-7 or T887) rice varieties employing genotyping-by-sequencing (Jang *et al.,* 2020). A total of 17 and 3 QTLs were detected for AAC and grain protein content, respectively. Allelic combination of qAAC6.1^{M23}and qAAC7.1^{T887}showed significantly higher content of associated amino acids (AAs) than other allelic combinations. In addition to AACs affected by qAAC7.1 in the primary QTL analysis, PC and other AACs, including Met, His,Lys, and Gly, were also significantly higher than those of both parents, indicating that qAAC7.1^{T887}could confer high AACs and PC consistently. QTLs associated with amino acid contents and linked/flanking markers are summarized in Table 2.

Table 1: QTLs reported for grain protein content in rice.

Population type	Cross	No. of QTLs	PVE range	Chromosomes number	Important QTL/Associated markers / genes	References
RILs	Zhenshan 97 (*Indica*) / Minghui 63 (*Indica*)	2	6.0-13.0	6,7		Tan et al. 2001
DH lines	Caiapo (*Indica*) / IRGC 103544 (*Oryza glaberrima*)	4	4.8-15.0	1,2,6,11	RM226-RM297	Aluko et al. 2004
DH lines	Gui630 (*Indica*) / 02428 (*Japonica*)	5	6.9-35.0	1,4,5,6,7		Hu et al. 2004
BIL (BC_3F_1)	v20A (*O. sativa*) / accession 103,544 (*O. glaberrima*)	1	9.0-10.0	8		Li et al. 2004
RILs	Moritawase (*Japonica*) / Koshihikari (*Japonica*)	3	2.3-16.3	2,6,9		Wada et al. 2006
BILs	Koshihikari (*Japonica*)/ Kasalath (*Indica*)// Koshihikari	2	14.3-14.8	6,10		Takeuchi et al. 2007
RILs	Chuan (*Indica*) / Nanyangzhan (*Japonica*)	2	2.69-4.50	6,7		Zhang et al. 2011
RILs	Xieqingzao B (*Indica*)/ Milyanga 46 (*Indica*)	5	3.9-19.3	3,4,5,6,10	RM251-RM282	Lou et al. 2009
RILs	Zhenshan97 (*Indica*) / Minghui 63 (*Indica*)	9	1.60-9.26	2,3,5,6,7,10,11,12		Yu et al. 2009
DH Lines	Samgang (*Tongli*) / Nagdong (*Japonica*)	3	6.92-22.98	1,11	RM287-RM26755	Shi et al. 2009
CSSLs	Asominori (*Japonica*) / IR24 (*Indica*)	9	3.0-53.7	1,2,3,6,8,11		Qin et al. 2009
RILs	Asominori (*Japonica*) / IR24 (*Indica*)	10	8.53-23.70	1,3,4,6,7,8,9,10,12		Zheng et al. 2011

RILs	Zhenshan97B (*Indica*) / Delong 208 (*Indica*)	2	7.2-25.9	1,7	RM445-RM418	Zhong et al. 2011
BILs	Koshihikari (*Japonica*)/ Kasalath (*Indica*)// Koshihikari	4	6.26-12.11	2,3,7,10		Zheng et al. 2012
DH Lines	Cheongcheong (*Indica*) / Nag-dong (*Indica*)	1	39-41	2	RM12532-RM555	Lee et al. 2014
DH Lines	CJ06 (*Japonica*) / TN1 (*Indica*)	1	12.3-15.8	10	RM216-RM467	Leng et al. 2014
DH Lines	Cheongcheong (*Indica*) / Nag-dong (*Indica*)	3	39-40	8,9,10	RM506-RM1235, RM24934-RM25128, RM219-RM23914	Yun et al. 2014
RILs	M201 / JY293	5	6.74-13.50	1,2,3,4	RM423-RM6375	Xu et al. 2015
CSSLs	Sasanishiki (*Japonica*) /Habataki (*Indica*)	1	10.38-15.43	1	RM7124	Yang et al. 2015
Association mapping	227 accessions	9	2, 4, 5, 6, 7, 8, 10, 11	5-20%	QTL (chr.11: 22240707 ~ 22563596)	Xu et al. 2016
ILs	Hwaseong (Jalonica)/ O. ru-fipogon (IRGC 105491)	5	2, 6, 11, 12			Yun et al. 2016
Association mapping	258 rice accessions	3	6.7%- 8.1%	2, 10	*qPC10.1* and *qPC10.2*, *qPC2*	Wang et al. 2017
DH Lines	Cheongcheong (*Indica*) / Nag-dong (*Indica*)	1	14	7	RM8261	Bruno et al 2017
BILs (BC_3F_5)	Naveen (*Indica*)/ ARC10075 (*Indica*)	3	6.70-17.35	1,2,7	CSCWR_ Os01g02590 _ 61041 (13.85), CSCWR_ Os02g10740 _65058 (6.70-17.35)	Chattopadhyay et al 2019

RIL	(HHZ (*Indica*) /JZ1560 (*Japonica*)	14	0.81%–18.59%	1, 2, 3, 4, 5, 7, 8, 10, 11	*qGPC1-1* delimited to a ~862 kb region flanked by JD1006 and JD1075	Wu et al. 2020
RIL	Yukihikari (*Japonica*)/ Joiku462 (*Japonica*)	8	22.3% 24.4%	1, 2, 3, 6, 8, 12	qPC6.2, qPC8	Kinoshita et al., 2017
CSSL	indica cultivar Habataki (*Indica*) / Sasanishiki (*Japonica*)	18	4.9-17.8%	1, 3, 6, 8, 9, 10, 11, 12	*qGPC-1* and *qGPC-10 (OsGluA2)*	Yang et al (2019)
BIL (BC3F3)	Kongyu 131/Cypress (population-I) and Kongyu 131/Vary Tarva Osla (population-II)	25 QTLs for total protein and protcin fraction	1.6% to 61.85%	All chromosomes	qGPC7.3 for total protein content validated	Alam et al. (2023)
Association panel	96 Diverse germplasm and improved lines	6		4, 9, 10, 12	*qGPC 4.1* and *qGPC 4.2, qGPC10.2,* RM 17,600 and RM 1272, RM 467	Chattopadhyay et al. 2023

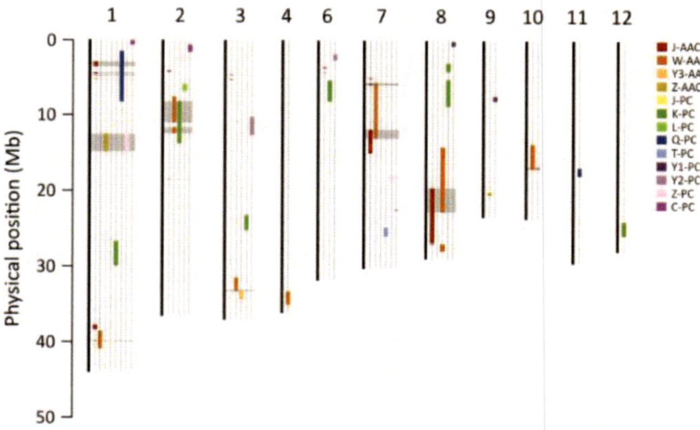

Fig. 4. Comparison of QTLs for grain protein (PC), protein fraction and amino acids content (AAC) distributed on all 12 chromosomes. Colour bars on chromosomes represent QTL region for AAC and PC reported by various studies. Gray blocks indicate the loci where two or more QTLs were co-localized. Each prefix of trait indicates the various studies (J, Jang *et al.*, 2020; C, Chattopadhyay *et al.*, 2019; K, Kinoshita *et al.*, 2017; L, Lou *et al.*, 2009; T, Tan *et al.*, 2001; W, Wang *et al.*, 2007; Y1, Yun *et al.*, 2014; Y2, Yu *et al.*, 2009; Y3, Yoo, 2007; Zhong *et al.*, 2011) (Reproduced from Jang *et al.* (2020), doi: 10.3389/fgene.2020.00240)

Table 2: QTLs detected for Amino acid content in rice

Population type		**No. of QTLs**	**PVE range (additive effect QTLs)**	**Chromosomes/Chromosome arms**	**Associated markers/ Genes**	**References**
RILs	Indica rice (Zhenshan 97) x Indica rice (Nanyangzhan)	2 QTL cluster	4.05-33.3	1,7	RM472-RM104	Wang *et al.*, 2007
RILs	Indica rice (Zhenshan 97) x Indica rice (Minghui 63)	10 (His)+ 8 (Arg)	12-35	1,2,3,5,6, 7,10,11,12		Zheng *et al.* (2008)
RILs	Indica rice (Zhenshan 97) x Indica rice (Minghui 63)	12	3.4-48.8	1,11		Zhong *et al.*, 2011
RILs	Indica rice (Zhenshan 97B) x Indica rice (Delong 208)	3 QTL cluster	4.2-31.7	1,7,9		Zhong *et al.*, 2011
RILs	O.sativa (Dasanbyeo) x O. sativa(TR22183)	QTL cluster of 6 aa (Ala, Val, Leu, Ile, and Phe)	10.2-12.4	3		Yoo, 2017
RIL	indica (Milyang 23 or M23) x japonica (Tong88-7 or T887)	A total of 17 QTLs			qAAC6. 1M23and qA-AC7.1T887	Jang *et al.*, 2020

References

Ali 2007. Effect of nitrate nitrite and ammonium, glutamine, glutamate 2-oxo glutarate on the RNA levels and enzyme activities of nitrite reductase and nitrate reductase in rice *physiol. Mol. Biol. Plants,* **13**(1): 17-25

Arima Y and Kumazawa K 1975 A kinetic study of amide and amino acid synthesis in rice seedling roots fed with^{15}N labelled ammonium (part 2). Physiological significance of glutamine on nitrogen absorption and assimilation in plants; *J. Sci. Soil. Manure. Jpn.* **46** 355–361.

Aluko, G., Martinez, C., Tohme, J., Castano, C., Bergman, C. and Oard, J. H., 2004. QTL mapping of grain quality traits from the interspecific cross *Oryza sativa*× *O. glaberrima. TheorAppl Genet.* **109**, 630.

Bloom AJ, Sukrapanna SS, Warner RL. 1992. Root Respiration Associated with Ammonium and Nitrate Absorption and Assimilation by Barley. *Plant Physiology*, **99**(4),1294–1301. https://doi.org/10.1104/pp.99.4.1294

Bassham, D. C., Laporte, M., Marty, F., Moriyasu, Y., Ohsumi, Y., Olsen, L. J., Yoshimoto, K., 2006. Autophagy in development and stress responses of plants. *Autophagy 2*(1), 2-11.

Beninati, N. F., Busch, R. H., 1992. Grain protein inheritance and nitrogen uptake and redistribution in a spring wheat cross. *Crop Sci.* **32**(6), 1471-1475.

Bao A, Zhao Z, Ding G, Shi L, Xu F, Cai H. 2014. Accumulated Expression Level of Cytosolic Glutamine Synthetase 1 Gene (OsGS1;1 or OsGS1;2) Alter Plant Development and the Carbon-Nitrogen Metabolic Status in Rice. *PLoS ONE*, **9**(4) e95581. https://doi.org/10.1371/journal.pone.0095581

Bao A, Liang Z, Zhao Z, Cai H. 2015. Overexpressing of OsAMT1-3, a High Affinity Ammonium Transporter Gene, Modifies Rice Growth and Carbon-Nitrogen Metabolic Status. *International Journal of Molecular Sciences*, **16**(12):9037–9063. https://doi.org/10.3390/ijms16059037

Bruno E, Yun-Sik Choi, Kyung Chung, Kyung–Min Kim. 2017. QTLs and analysis of thecandidate gene for amylose, protein, and moisture content in rice (*Oryza sativa* L.).*3 Biotech*, 7:40, DOI 10.1007/s13205-017-0687-8

Chattopadhyay K, Behera L, Bagchi TB, Sardar SS, Moharana N, Patra NR, Chakraborti M, Das A, Marndi BC, Sarkar A, Umakanta N, Chakraborty K, Bose LK, Sarkar S, Ray S, Sharma SG. 2019a. Detection of stable QTLs for grain protein content in rice (Oryza sativa L.) employing high throughput phenotyping and genotyping platforms.*Sci Rep*, 9: 3196, https://doi.org/10.1038/s41598-019-39863-2.

Cheng-gang LIANG, Li-ping CHEN, Yan WANG, Jia LIU, Guang-li XU, Tian LI. 2021. High Temperature at Grain-filling Stage Affects Nitrogen Metabolism Enzyme Activities in Grains and Grain Nutritional Quality in Rice Rice Science, **18**(3):210-216, https://doi.org/10.1016/S1672-6308(11)60029-2

Cai H, Zhou Y, Xiao J, Li X, Zhang Q, Lian X. 2009. Overexpressed glutamine synthetase gene modifies nitrogen metabolism and abiotic stress responses in rice. *Plant Cell Reports*, **28**(3): 527–537. https://doi.org/10.1007/s00299-008-0665-z

Diaz, C., Saliba-Colombani, V., Loudet, O., Belluomo, P., Moreau, L., Daniel-Vedele, F., ...Masclaux-Daubresse, C., 2006. Leaf yellowing and anthocyanin accumulation are two genetically independent strategies in response to nitrogen limitation in Arabidopsis thaliana. *Plant Cell Physiol.* **47**(1), 74-83.

Duan Y, Zhang Y, Ye LT, Fan X, Xu GF, Shen Q. 2007. Responses of Rice Cultivars with Different Nitrogen Use Efficiency to Partial Nitrate Nutrition. *Annals of Botany*, **99**(6): 1153–1160. https://doi.org/10.1093/aob/mcm051

Duan M, Sun SSM. 2005, September. Profiling the Expression of Genes Controlling Rice Grain Quality. *Plant Molecular Biology*, **59**(1): 165–178. https://doi.org/10.1007/s11103-004-7507-3

Friedman, M., Brandon, D. L., Nutritional and health benefits of soy proteins. *J.Agric. Food Chem*. 2001, **49**, 1069–1086.

Epstein E, Bloom AJ (2005) Mineral Nutrition of Plants: Principles and Perspectives. 2nd Edition. Sinauer Associates, Sunderland, MA, 405 pp.

Fageria, NK, Carvalho M, Dos Santos FC. 2014. Root growth of upland rice genotypes as influenced by nitrogen fertilization J. Plant Nutr., **37**: 95-106

Fischer, W. N., Loo, D. D., Ludewig, U., Boorer, K. J., Tegeder, M., Rentsch, D., Frommer, W. B., 2002. Low and high affinity amino acid H+co-transporters for cellular import of neutral and charged amino acids. *The Plant J.* **29**(6), 717-731.

Forde, B. G., Lea, P. J., 2007. Glutamate in plants: metabolism, regulation, and signalling. *J. experimental bot.* **58**(9), 2339-2358.

Gaufichon, L., Masclaux-Daubresse, C., Tcherkez, G., Reisdorf-Cren, M., Sakakibara, Y., Hase, T., *et al.,* (2013). Arabidopsis thaliana ASN2 encoding asparagine synthetase is involved in the control of nitrogen assimilation and export during vegetative growth. Plant Cell Environ. **36**: 328–342. doi: 10.1111/j.1365-3040.2012.02576.x

Gaur VS, Singh U, Gupta A, Kumar A. 2012. Influence of different nitrogen inputs on the members of ammonium transporter and glutamine synthetase genes in two rice genotypes having differential responsiveness to nitrogen. *Molecular Biology Reports*, **39**(8):8035–8044. https://doi.org/10.1007/s11033-012-1650-8

Galangau, F., Daniel-Vedele, F., Moureaux, T., *et al.,* (1988) Expression of Leaf Nitrate Reductase Gene from Tomato and Tobacco in Relation to Light-Dark Regimes and Nitrate Supply. Plant Physiology, 88, 383-388. https://doi.org/10.1104/pp.88.2.383

Guiz, C., Hirel, B., Shedlofsky, G., and Gadal, P. (1979). Occurrence and influence of light on the relative proportions of two glutamine synthetases in rice leaves. *Plant Sci. Lett.* **15**: 271–277.

Hille. R. 1996. The mononuclear molybdenum enzymes. Chemical Reviews, 96, pp. 2757-2816

Hayakawa, T.,Nakamura, T.,Hattori,F.,Mae, T.,Ojima,K.,andYamaya, T. (1994). Cellular-localization of NADH-dependent glutamate-synthase protein in vascular bundles of unexpanded leaf blades and young grains of rice plants. *Planta* **193**:455–460.

Hirner, A., Ladwig, F., Stransky, H., Okumoto, S., Keinath, M., Harms, A., Koch, W., 2006. Arabidopsis LHT1 is a high-affinity transporter for cellular amino acid uptake in both root epidermis and leaf mesophyll. *The Plant Cell* **18**(8), 1931-1946.

Hirel, B., Bertin, P., Quilleré, I., Bourdoncle, W., Attagnant, C., Dellay, C., ...Gallais, A., 2001. Towards a better understanding of the genetic and physiological basis for nitrogen use efficiency in maize. *Plant Physiol.* **125**(3): 1258-1270.

Hu ZL, Li P, Zhou MQ, Zhang ZH, Wang LX, Zhu LH, *et al.,* Mapping of quantitative trait loci (QTLs) for rice protein and fat content using doubled haploid lines. Euphytica. 2004;135:47

Islam MS. 2019. Sensing and Uptake of Nitrogen in Rice Plant: A Molecular View. *Rice Science*, **26**(6): 343-355

Ishiyama, K., Inoue, E., Watanabe-Takahashi, A., Obara, M., Yamaya, T., Takahashi, H., 2004. Kinetic properties and ammonium-dependent regulation of cytosolic isoenzymes of glutamine synthetase in Arabidopsis. *Journal of Biological Chemistry,* **279**(16), 16598-16605.

James D, Borphukan B, Fartyal D, Ram B, Singh J, Manna M, Sheri V, Panditi V, Yadav R, Achary VMM, Reddy MK. 2018. Concurrent Overexpression of OsGS1;1 and OsGS2 Genes in Transgenic Rice (*Oryza sativa* L.): Impact on Tolerance to Abiotic Stresses. *Frontiers in Plant Science*, **9**. https://doi.org/10.3389/fpls.2018.00786

Jang S, Han J-H, Lee YK, Shin N-H, Kang YJ, Kim C-K and Chin JH (2020) Mapping and Validation of QTLs for the Amino Acid and Total Protein Content in Brown *Rice. Front. Genet.* **11**:240. doi: 10.3389/fgene.2020.00240

Kamachi, K., Yamaya, T, Mae, T., Ojima K. 1991. A role for glutamine synthetase in the remobilization of leaf nitrogen during natural senescence in rice leaves, *Plant Physiol*, **96** (2) (1991), pp. 411-417

Kaur, G., B. Asthir, N.S. Bains and M. Farooq, 2015. Nitrogen nutrition, its assimilation and remobilization in diverse wheat genotypes. *Int. J. Agric. Biol.*, **17**: 531-538

Kaur, A., Shevkani K, Katyal M, Singh N, Ahlawat AK, Singh AM. 2016. Physicochemical and rheological properties of starch and flour from different durum wheat varieties and their relationships with noodle quality, *Journal of Food Science and Technology*, **53**(4) (2016), pp. 2127-2138

Kichey, T., Hirel, B., Heumez, E., Dubois, F., Le Gouis, J., 2007. In winter wheat (*Triticum aestivum* L.), post-anthesis nitrogen uptake and remobilisation to the grain correlates with agronomic traits and nitrogen physiological markers. *Field Crops Res.* **102**(1):22-32.

Kawakatsu, T., Yamamoto, M.P., Hirose, S., Yano, M. and Takaiwa, F. (2008) Characterization of a new rice glutelin gene GluD-1 expressed in the starchy endosperm. *J. Exp. Bot.*, **59**, 4233–4245.

Kinoshita, N., Kato, M., Koyasaki, K., Kawashima, T., Nishimura, T., Hirayama, Y., *et al.*, 2017. Identification of quantitative trait loci for rice grain quality and yield-related traits in two closely related Oryza sativa L. subsp. Japonica cultivars grown near the northernmost limit for rice paddy cultivation. *Breed. Sci.* **67**,191–206. doi: 10.1270/jsbbs.16155

Kubota, M. *et al.*, 2010. Improvement in the in vivo digestibility of rice protein by alkali extraction is due to structural changes in prolamin/protein body–I particle. *Bioscience Biotechnology and Biochemistry*, **74**, 614-619.

Katsube-Tanaka, T., Duldulao, J.B.A., Kimura, Y., Iida, S., Yamaguchi, T., Nakano, J., Utsumi, S., 2004. The two subfamilies of rice glutelin differ in both primary and higher-order structures. *Biochimica et Biophysica Acta-Proteins and Proteomics* **1699**, 95e102.

Kumar A., Silim S. N., Okamoto M., Siddiqi M. Y., Glass A. D. (2003). Differential expression of three members of the AMT1 gene family encoding putative high-affinity NH4+ transporters in roots of *Oryza sativa* subspecies indica. *Plant Cell Environ.* **26**, 907–914. 10.1046/j.1365-3040.2003.01023.x

Leng Y, Xue D, Yang Y, Hu S, Su Y, Huang L, *et al.*, 2014. Mapping of QTLs for eating and cooking quality-related traits in rice (Oryza sativa L.). *Euphytica*, **197**:99

Lee GH, Yun BW, Kim KM. 2014. Analysis of QTLs associated with the rice quality related gene by double haploid populations. International Journal of Genomics.

Li J, Xiao J, Grandillo S, Jiang L, Wan Y, Deng Q, *et al.*, QTL detection for rice grain quality traits using an inter-specific backcross population derived from cultivated Asian (*O. sativa* L.) and African (*O. glaberrima* S.) rice. Genome. 2004;47:697

Lou J, Chen L, Yue G, Lou Q, Mei H, Xiong L, *et al.*, 2009. QTL mapping of grain quality traits in rice. *Journal of Cereal Science*. **50**:145

Lee, M., Huang, T., Toro-Ramos, T., Fraga, M., Last, R. L., and Jander,G. (2008). Reduced activity of Arabidopsis thaliana HMT2, a methioninebiosynthetic enzyme, increases seed methionine content. *Plant J.* **54**, 310–320.doi: 10.1111/j.1365-313X.2008.03419.x

Lea, P. J., Sodek, L., Parry, M. A., Shewry, P. R., and Halford,N. G. (2007). Asparagine in plants. Ann. App. Biol. 150, 1–26.doi: 10.1111/j.1744-7348.2006.00104.x.

Lea, P. J. and Miflin, B. J. (2011)."Nitrogen assimilation and its relevanceto crop improvement," in Nitrogen Metabolism in Plants in the Post-Genomic Era. *Annual Plant Reviews*, **42** (Oxford: Wiley-Blackwell), 1–40.doi: 10.1002/9781119312994.apr0448

Moison, M., Marmagne, A., Dinant, S., Soulay, F., Azzopardi, M., Lothier, J., *et al.,* (2018). Three cytosolic glutamine synthetase isoforms located in different order veins work together for N remobilization and seed filling in arabidopsis. *J. Exp. Bot.* **69**: 4379–4393. doi: 10.1093/jxb/ery217

Martin, A., Lee, J., Kichey, T., Gerentes, D., Zivy, M., Tatout, C. and Tercé-Laforgue, T., 2006. Two cytosolic glutamine synthetase isoforms of maize are specifically involved in the control of grain production. *The Plant Cell*, **18**(11), 3252-3274.

Mitsukawa, N., Hayashi, H., Yamamoto, K., Kidzu, K., Konishi, R., Masumura, T. and Tanaka, K. 1998. Molecular cloning of a novel glutelin cDNA from rice seeds. *Plant Biotechnol.*, **15**, 205–211.

Patrick, J. W., Offler, C. E., 2001. Compartmentation of transport and transfer events in developing seeds. *J. Experimental Bot.* **52**(356): 551-564.

Peng, B., Kong, H., Li, Y., Wang, L., Zhong, M., Sun, L., *et al.,* 2014. OsAAP6 functions as an important regulator of grain protein content and nutritional quality in rice. *Nat. Commun.* **5:** 4847.

Panthee, D. R., Pantalone, V. R., Saxton, A. M., West, D. R. and Sams, C. E., 2006. Genomic regions associated with amino acid composition in soybean. *Mol. Breed.* , **17**: 79–89.

Pradhan SK, Pandit E, Pawar S, Bharati B, Chatopadhyay K, Singh S, Dash P, Reddy JN. 2019. Association mapping reveals multiple QTLs for grain protein content in rice useful for biofortification. *Molecular Genetics and Genomics*, **294**(4):963–983. https://doi.org/10.1007/s00438-019-01556-w

Qin Y, Kim SM. Sohn JK. 2009. QTL analysis of protein content in double-haploid lines of rice. *Korean Journal of Crop Science*. **54:**165

Reggiani R, Nebuloni M, Mattana M, Brambilla I. 2000. Anaerobic accumulation of amino acids in rice roots: role of the glutamine synthetase/glutamate synthase cycle. *Amino Acids* **18**: 207–217

Rentsch, D., Schmidt, S., Tegeder, M., 2007. Transporters for uptake and allocation of organic nitrogen compounds in plants. *FEBS letters* **581**(12):2281-2289.

Shi CH, Shi Y, Lou XY, Xu HM, Zheng X, Wu JG. 2009. Identification of endosperm and maternal plant QTLs for protein and lysine contents of rice across different environments. *Crop & Pasture Science*. **60**:295

Suzuki, A., and Gadal, P. (1982). Glutamate synthase from rice leaves. *Plant Physiol*. **69**:848–852.

Tan, Y. F., Sun, M., Xing, Y. Z., Hua, J. P., Sun, X. L., Zhang, Q. F. and Corke, H., 2001. Mapping quantitative trait loci for milling quality, protein content and color characteristics of rice using a recombinant inbred line population derived from an elite rice hybrid. *TheorAppl Genet.,* **103**, 1037.

Takaiwa, F., Oono, K. and Kato, A. 1991. Analysis of the 5¢ flanking region responsible for the endosperm-specific expression of a rice glutelin chimeric gene in transgenic tobacco. *Plant Mol. Biol*., **16:**49–58.

Takeuchi Y, Nonoue Y, Ebitani T, Suzuki K, Aoki N, Sato H, *et al.,* 2007. QTL detection for eating quality including glossiness stickiness taste and hardness of cooked rice. *Breeding Science*. **57**:231

Tabuchi, M., Sugiyama, K., Ishiyama, K., Inoue, E., Sato, T., Takahashi, H., & Yamaya, T. 2005. Severe reduction in growth rate and grain filling of rice mutants lacking OsGS1;1, a cytosolic glutamine synthetase1;1. *The Plant Journal*, **42**(5):641–651. https://doi.org/10.1111/j.1365-313x.2005.02406.x

Wada T, Uchimura Y, Ogata T, Tsubone M, Matsue Y. 2006. Mapping of QTLs for physicochemical properties in japonica rice. *Breeding Science*. **56**:253

Wang, L., Zhong, M., Li, X., Yuan, D., Xu, Y., Liu, H., *et al.,* (2007). The QTL controlling amino acid content in grains of rice (*Oryza sativa*) are co-localized with the regions involved in the amino acid metabolism pathway. *Mol. Breed.* **21:**127–137. doi: 10.1007/s11032-007-9141-7

Wang, L., Zhong, M., Li, X., Yuan, D., Xu, Y., Liu, H., He, Y., Luo, L. and Zhang, Q., 2008. The QTL controlling amino acid content in grains of rice (*Oryza sativa*) arecolocalized with the regions involved in the amino acid metabolism pathway. *Mol Breed.* **21**, 127.

Wu YB, Li G, Zhu YJ, Cheng YC, Yang JY, Chen HZ, Song XJ, Ying JZ. 2020. Genome-Wide Identification of QTLs for Grain Protein Content Based on Genotyping-by-Resequencing and Verification of qGPC1-1 in Rice. *International Journal of Molecular Sciences*, **21**(2): 408. https://doi.org/10.3390/ijms21020408

Xu F, Sun C, Huang Y, Chen Y, Tong C, Bao J. 2015. QTL mapping for rice grain quality: A strategy to detect more QTLs within sub-populations. *Molecular Breeding*. **35:**105

Yang, Y., Guo, M., Li, R., Shen, L., Wang, W., Liu, M., Zhu, Q., Hu, Z., He, Q., Xue, Y. and Tang, S., 2015. Identification of quantitative trait loci responsible for rice grain protein content using chromosome segment substitution lines and fine mapping of qPC-1 in rice (*Oryza sativa* L.). *Mol Breed.,***35,** 130.

Yoo, S.-C. (2017). Quantitative trait loci controlling the amino acid content in rice (*Oryza sativa* L.). *J. Plant Biotechno.*, **44**: 349–355. doi: 10.5010/JPB.2017.44.4.349

Yu YH, Li G, Fan YY, Zhang KQ, Min J, Zhu ZW, *et al.,* 2009. Genetic relationship between grain yield and the contents of protein and fat in a recombinant inbred population of rice. *Journal of Cereal Science*. **50**:121

Yun BW, Kim MG, Handoyo T, Kim KM. 2014. Analysis of rice grain quality-associated quantitative trait loci by using genetic mapping. *Asian Journal of Plant Sciences*. **5**(09):1125

Yang Y, Guo M, Li R, Shen L, Wang W, Liu M, *et al.,* 2015. Identification of quantitative trait loci responsible for rice grain protein content using chromosome segment substitution lines and fine mapping of qPC-1 in rice (*Oryza sativa* L.). *Molecular Breeding*. **35**:130

Zeng, D. D., Qin, R., Li, M., Alamin, M., Jin, X. L., Liu, Y., *et al.,* 2017. The ferredoxin-dependent glutamate synthase (OsFd-GOGAT) participates in leaf senescence and the nitrogen remobilization in rice. *Mol. Genet. Genomics* **292:**385–395. doi: 10.1007/s00438-016-1275-z

Zheng, L., Zhang, W., Liu, S., Chen, L., Liu, X., Chen, X., Ma, J., Chen, W., Zhao, Z., Jiang, L. and Wan, J., 2012. Genetic relationship between grain chalkiness, protein content, and paste viscosity properties in a backcross inbred population of rice. *J Cereal Sci.* 2012, **56:** 153.

Zhong, M., Wang, L. Q., Yuan, D. J., Luo, L. J., Xu, C. G. and He, Y. Q., 2011. Identification of QTL affecting protein and amino acid contents in rice.*Rice Science.,* **18:** 187.

Zheng, X., Jian-Guo, W. U., Xiang-Yang, LOU., Hai-Ming, XU., and Chun-Hai, SHI., QTL 2008. Analysis of maternal and endosperm genomes for histidine and arginine in rice (*Oryza sativa* L.) across environments. *ActaAgronomica Sinica.***34:** 369.

Zheng, L., Zhang, W., Chen, X., Ma, J., Chen, W., Zhao, Z., Zhai, H. and Wan, J., 2011. Dynamic QTL analysis of rice protein content and protein index using recombinant inbred lines. *J Plant Biol.* **54:**321.

Zheng L, Zhang W, Chen X, Ma J, Chen W, Zhao Z, *et al.,* 2011. Dynamic QTL analysis of rice protein content and protein index using recombinant inbred lines. *Journal of Plant Biology*. **54:**321

Zhang X, Zhang G, Guo L, Wang H, Zeng D, Dong G, *et al.,* 2011. Identification of quantitative trait loci for Cd and Zn concentrations of brown rice grown in Cd-polluted soils. *Euphytica*. **180:**173

3

Variation in Grain Protein and Essential Amino Acid Content and Their Role in the Nutritional Improvement of Rice

Introduction

Understanding the variation in grain protein content is the pre-requisite of its improvement in modern rice varieties. The measure of quantitative and qualitative aspects of grain storage protein content includes estimation of soluble and total grain protein, quantifying various fractions, viz. glutelins, prolamins, globulins and albumins and their profiling through SDS-Page and estimation of amino acid. As traditional chemical methods for protein content determination are destructive in nature. All these estimates comprehensively give insight into the diversity of storage protein in rice germplasm. The standard procedures followed for all these estimates are briefly stated in the following sections.

Estimation of Total Grain Protein Content in Rice

The total protein is estimated via estimation of total N through the Kjeldahl method whereas soluble protein is estimated through Lowry method, Dumas method and Bradford method. But for estimation of large number of genotypes all these processes are tedious, time-consuming, and costly. Near-infrared spectroscopy (NIRS), as a non-destructive detection technique, can achieve rapid quantitative determination of protein contents and become an important development direction to replace traditional chemical detection methods.

a. Through micro-Kjeldahl method

Grain protein content is determined by the standard micro-Kjeldahl method (Yoshida *et al.,* 1976). Around 200 mg of brown rice, that is grains from which the husk is removed but the brown layer of bran is retained (which is normally removed during polishing), from each sample are taken in a Kjeldahl flask in two replications. After digestion with H_2SO_4 the final solution is titrated with 0.01N HCl and the reading taken. The grain protein content (GPC) is calculated using the formula: GPC (%) = (X × 0.09 × 14 × 100/0.2 × 1000) × 5.95 where 'X' is titrated value. Single grain protein content was measured as the average of the 10 grains and expressed as mg/g

b. Calibration of near infrared spectroscopy (NIRS) and estimation of apparent GPC

Near-infrared (NIR) spectroscopy has proved effective in predicting GPC and is a reliable tool for genetical analysis and high-throughput selection (Shao *et al* 2011). This is based on the absorption of molecular overtone and combination vibrations of hydrogenous groups X-H (X =C, N, O) in the near-infrared region of the electromagnetic spectrum. The wavelength range is from 750 nm to 2500 nm. The calibration model is developed and used to replace the wet chemistry analysis. Several applications of NIR spectroscopy to predict grain protein in rice samples are available (xu *et al.,* 2019).

Bagchi *et al.* (2015) using more than 160 diverse rice genotypes with a wide range of variation in grain protein content calibrated the NIR spectroscopy by taking the best equation based on lowest SEC (Standard error of calibration) and SECV (Standard error of cross validation) and highest 1-VR (1 minus variance ratio) and RSQ (Coefficient of determination). Before NIR analysis, the samples were kept at room temperature (25°C) for 6 h to balance the moisture and temperature as these factors can affect the reflectance and absorbance of NIR wave. The NIRS was calibrated with the help of three softwares related to NIR spectroscopy. A cup was used for scanning of the sample with full spectrum (400-2500 nm) taking about 15 g of each of 160 samples. The reflectance spectra (log1/R) from 400-2500 nm were recorded at 2 nm intervals (Fig 1). A subsequent external validation of the initial calibration model using samples independent from the calibration set led to further NIRS performance values for each constituent.1,4,4,1 and 1,4,3,1 both were good calibration equation based on Lowest SEC,SECV and SEP and highest RSQ, 1-VR (Bagchi *et al.,* 2015).

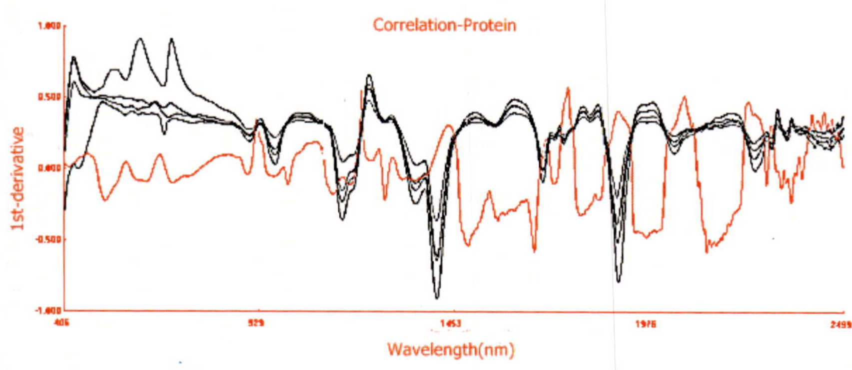

Fig. 1: Regression coefficient along with correlation plot of modified partial least squares model (MPLS) calibration equations under first derivative for grain protein content (Bagchi *et al.*, 2016; http://dx.doi.org/10.1016/j.foodchem.2015.05.038)

Extraction, Fractionation and Quantification of Storage Proteins

Extraction of rice proteins is generally performed based on procedure described by Ju *et al.* (2001) and its modification (Krishnan and Okita, 1986). Rice flour (6-7 g) is defatted with n-hexane. The defatted flour (5g) is extracted by stirring in 20 mL of distilled water at room temperature (RT) for 2 hrs to g*et al*bumin fraction. After centrifugation at 3,000×g for 30 min the residue is re-extracted with 20 mL of 5% NaCl for 2 hrs at room temperature to obtain the globulin fraction. This is followed by centrifugation at 3,000×g for 30 min and the residue is extracted for prolamins with 15 mL of 70% ethanol for 2 hrs followed by glutelin extraction with 37.5 mL of 0.2 M sodium borate buffer (pH10) containing 0.5% sodium dodecyl sulphate (SDS) and 0.6% β-mercaptoethanol at RT for 2 hrs (Juliano, 1980; Sugimoto, 1986). Each extraction is generally repeated for two times. The glutelins and prolamins from rice flour are extracted and partially purified essentially following the procedure described by Krishnan and Okita (1986).The extracted proteins are freeze-dried and stored at -70°C. The protein content of each fraction is measured according to the method stated by Lowry *et al.* (1951).

Developing Storage Protein Profile using SDS-PAGE Analysis

SDS-PAGE gels is used to fraction the partially purified glutelins and prolamines. After dissolving in extraction buffer, glutelin solutions are mixed with the same volume of 2x SDS sample buffer (100 mMTris–HCl buffer (pH 6.8) 4% SDS, 20% glycerol, 10% 2-mercaptoethanol and 0.02% bromophenol blue). SDS-PAGE is carried out according to the standard protocol described by Sambrook and Russel (2001) with 5% stacking and

12% resolving Polyacrylamide gel. After electrophoresis peptide pattern in the gel is visualized after staining with Coomassie brilliant blue followed by photography in a gel documentation system.

Amino Acid Profiling of Rice Grain

Grinded rice samples are hydrolyzed with 6N hydrochloric acid at 110°C under anaerobic condition for 24h and the hydrolyzed samples are neutralized with 6N NaOH. Finally these are derivatized using a kit and are injected in high performance liquid chromatography (HPLC) equipped with a fluorescence detector. The amino acids are identified and quantified by comparing with the retention times and peak areas of standards following a procedure described by Pal *et al.* (2016).

Diversity of Rice Germplasms for GPC

Many reports are available on the variability for grain protein content of rice germplasm with diverse nature and from various geographical locations. International Rice Research Institute (IRRI) evaluated 13089 *indica* accessions. Protein content of those lines was ranged from 4.3% to 18.2% (dry season) and 3.5% to 15.9% (wet season). This observation indicated wide genetic variability of this quantitative trait. Juliono and Villareal (1993) reported that protein content in 1622 milled rice samples derived from 24 countries ranged from 4 to 14 percent and mean protein ranged from 6.3 to 9.2 percent and the overall mean was 7.8 percent. Kennedy and Burlingame (2003) analyzed the protein contents of 2,674 samples of *O. sativa* accessions and found 8.8% as the mean, ranging from 4.5 to 15.9%. Chandel *et al.* (2005) also observed a wide variation ranging from 2.8% to 9.9% for protein concentration in milled rice germplasm of Chhattisgarh. Cao *et al.* (2009) reported a wide variation in the levels of storage protein content (7.38–15.41%) in Chinese varieties of *O. sativa* rice. De Silveira *et al.* (2010) found a wide range (4.4– 20.2%) with an average of 10.31% for grain protein contents of 550 accessions in the rice Core Collection of Embrapa. Among 258 land races of rice belonging to the extra-early group (95 days) maintained at IGKV, Raipur, Chhattisgarh, Banerjee *et al.* (2011) found a wide variation of protein content in milled rice samples which ranged from 4.91% -12.08% with the mean of 6.63%. Rice germplasm lines of Chhattisgarh core collection for grain protein content were also analyzed and high protein germplasm were identified (CGR-436, 11.2%); GP-145-48, 10.68%; CGR-446, 10.43%; CGR-52, 10.35%; CGR-77, 9.92%) (Patil *et al.* 2014). Chattopadhyay *et al.* (2011) assessed diversity of grain protein of rice germplasm collected from North-eastern India. At ICAR-

NRRI, Cuttack Mohanty *et al.* (2011) reported 16.41% and 15.27% protein in brown rice of rice accessions, ARC 10063 and ARC 10075. Respectively. In this Institute, around 2000 germplasm of *Indica* rice were evaluated for protein content. A wide range of variation was observed ranging from 4.01% (ARC 5973) to 14% (PB -312). Rice cultivars, Kalinga-III, Heera, Mamihanger, Bindli with an average 11-13% protein content were and used as donors in the rice breeding programme (Chattopadhyay *et al.,* 2018). Estimation of protein content in 150 rice germplasm accessions was done by Aiyswaraya *et al.* (2017) to identify protein rich germplasm such as Mapillaisamba, Kudaivazhai and Norungan with very high protein content in the range of 12.38 to 14.54 g/100g rice. Santos *et al.* (2013) analyzed twenty nine accessions of the wild rice species *Oryza glumaepatula,* collected from five Brazilian states and two commercial cultivars for storage protein profile and amino acid content. Total protein levels ranged from 14.94% (wild genotype BGA14280) to 9.07% (BGA14179).

Characterization of Rice Protein and Variability in Protein Quality

The distribution of protein in rice is extremely uneven. Zhao *et al.* (2016) stained rice slices with Coomassie brilliant blue R250 and confirmed that storage protein are mainly distributed in the aleurone layer and embryo. In terms of protein types, albumin and globulin are more abundant in the outer tissues such as the aleurone layer and glume, while glutelin is the most abundant (83%) in endosperm, and prolamin is evenly distributed in rice bran, fine bran and milled rice (Juliano, 1993; Ren *et al.*, 2002). Accordingly, milled rice contains lower quantities of protein than brown rice because, during milling, a part of the protein-rich outer layers (aleurone cells) is removed (Table 1).

Table 1: Proximate protein contents of rough rice and its milling fractions (Juliano, 1993).

Rice	Crude protein content (g)
Brown rice	7.1–8.3
Milled rice	6.3–7.1
Rice bran	11.3–14.9
Rice hull	2.0–2.8

Kim *et al.* (2013) found that waxy brown rice contained 1.86, 0.50, 7.31 and 0.05% albumin, globulin, glutelin and prolamin fractions, respectively. Chattopadhyay *et al.* (2018) reported the values of albumin, globulin, glutelin and prolamin among the selected breeding lines ranged from 0.15 to 0.26%, 0.18 to 0.42%, 2.72 to 7.05% and 0.08 to 0.11%, respectively. They also detected the basic sub-unit, α-glutelin in an average 29 kD region, while the average molecular weight of β-glutelin unit was 21 kD in the studied genotypes. Another prolamin band was also observed in those genotypes at 13-14 kD region. The SDS-PAGE profile in that study was found to be similar to the banding pattern observed by Wen & Luthe (1985), But other researchers (Mahmoud *et al.* 2008; Jiang *et al.* 2014; Pal *et al.* 2016) obtained bands with slightly higher molecular weights for α-glutelin. Two high-protein rice germplasm, ARC 10063 and ARC 10075, identified from the stock of the Asom Rice Collections of the NRRI (CRRI) Rice Gene Bank, were evaluated thoroughly (Mohanty *et al.* 2011). Three glutelin bands (near 21 kD, 29 kD and 43 kD) are highly expressed in the high-protein cultivars. They showed higher activity of nitrate reductase (NR) and glutamic dehydrogenase (GDH) at seedling stage (one-week-old) and maximum tillering stage (three week-old). Minatik Charang (ARC 10075) was registered by the Plant Germplasm Registration Committee. Nitrate reductase (NR) and Nitrite reductase (NiR) activity was found to be higher for ARC 10075 as compared to low protein counterpart. It was also found with higher free amino acids (0.26%), methionine (0.082%), Lysine (0.049%) and Tryptophan (0.063%) (Mohanty *et al.,* 2011). The fractions of different soluble protein in this germplasm are 0.434%, 1.415%, 0.443% and 12.864% of Albumins, globulins, prolamins and glutelins, respectively (Chattopadhyay *et al*., 2019a). The yield of this germplasm was 2.45 t/ha. The average maturity duration was found medium (130 days). It has tall (140cm) plant type and The grain type is medium slender. It has having high head rice recovery (61%), high Alkali spreading value (7) and medium amylose content (24.6%) (ICAR-NRRI Annual Report 2013-14).

Breeding for High Protein Rice

Strong efforts have been made over the past few decades to improve the protein content and quality of rice through conventional breeding and induced mutagenesis (Mahmoud *et al.,* 2008; Khush and Juliono 1984). However, these efforts have been largely unsuccessful, as indicated by the lack of modern high-protein rice cultivars (Vasal, 2002). Various methodologies were employed in improvement of rice for grain protein content. In earlier rice breeding programme for developing high protein elite lines at IRRI, pedigree and long cycle recurrent selection were followed and one elite line IR480 was identified with a higher percentage of grain protein than IR8 and same grain yield as IR8. But developed high protein lines were not accepted in long run probably either due to deviation in grain type and cooking qualities from the adapted parent, IR 8, or due to low stability of their protein yield. Mahmoud *et al.* (2008) found a significant increase in seed protein content in an interspecific hybrid between *Oryza sativa* ssp. *indica* and the wild species *Oryza nivara*. In the last decade, some advances have been made in rice protein quality breeding, especially in glutelin (Hua *et al*., 2012;Wong *et al*., 2015; Chattopadhyay *et al*., 2018; Lee *et al*., 2021).

High protein rice through pedigree and Bulk-pedigree method of breeding

As GPC is a polygenic trait (Mahmoud *et al.,* 2008), it generally shows low heritability. Tsuzuki and Kuroda (1989) developed three breeding lines (F_8) of rice, derived from cross between Brimful X Koshihikari which were tested for protein, amino acid, growth test with rats and yield parameters in field trials. The breeding lines had higher protein and amino acid concentrations than the leading Japanese cultivar Koshihikari, which was grown as a control. Grain yield in the breeding lines was lower than that of Koshihikari (4.5 t/ha). The grain-protein yield of the breeding lines was higher (500 kg/ha) than that of Koshihikari (440 kg/ha). ICAR-NRRI, Cuttack has developed a few high yielding breeding lines with substantial improvement of grain protein content over the high yielding varieties through bulk-pedigree breeding methods (Chattopadhyay *et al*., 2018). Significant improvement in glutelin content in transgressive segregants were found. The bulk-pedigree method is generally used for improving traits with low heritability. This is a modification of the bulk method, in which individual plants of the F2 generation are harvested in bulk up to the F4 generation followed by single-plant selections in subsequent generations which is similar to the pedigree method. Wynne and Gregory (1981), who pioneered this modification, observed that selection in advanced generation through the pedigree method would facilitate in fixing the desirable genes. Although there is no direct effect of GPC on seed yield, the marginally negative correlation between the two traits revealed the possible risk of yield-penalty when selection is confined solely to higher GPC. Therefore, Chattopadhyay *et al.* (2018) opted the pedigree selection cycle for simultaneous selection for higher GPC and a higher number of panicles per plant, being the most important component of grain yield revealed through association analysis and further supported by PCA. The negative association was diminished to a significant level after three pedigree selection cycles and this expedited the simultaneous improvement GPC and grain yield. Three high-yielding Indica rice cultivars (IR64, Swarna and Naveen) and another Indica cultivar (Sharbati) with good grain quality and high iron (Fe) and zinc (Zn) contents were crossed with two Indica rice donors for high GPC (ARC10075 and ARC10063) (Chattopadhyay *et al.,* 2018). Around 2500 F_2 plants per population were raised. From 200 F2 plant progenies, 42 were selected based on desirable plant type and higher yield potential from five different cross combinations (IR64 × ARC10075, IR64 × ARC10063, Swarna × ARC10063, Naveen × ARC10063 and Sharbati × ARC10063). Through bulk-pedigree method 1710 F8 selection lines were derived in total from five breeding populations (Fig. 2). For GPC, the selection differential (S) was significant for all the crosses and for the population as a whole. On the other hand, S for protein yield was significant for all the crosses and for the population as a whole and, as expected, R was significant for the population as a whole. They found that observed and expected R for GPC and protein yield were not significantly different. This finding validated the method adopted in that breeding programme and proved that the bulk-pedigree method was effective on traits significantly influenced by genotype x environment interaction. As suggested by Frey and Horner (1955), a close agreement between the expected and actual gains indicates that the gene action involved in the selection is largely additive.

Higher expression in both α- and β-glutelin was observed in most of the high protein lines than the high yielding cultivar Swarna (Fig. 3). This banding pattern mostly correlated with the glutelin content observed through fractionation of soluble protein. The SDS-PAGE profile of genotypes was found to be similar to the banding pattern obtained by Wen & Luthe (1985), where an α-glutelin polypeptide group with three bands had an average molecular weight of 29.6 kD and the smaller group, β-glutelin with two bands, had an average molecular weight of 21.1 kD. The higher intensity of both groups of bands in high-protein lines compared with Swarna indicates that the enhanced GPC is due primarily to higher accumulation of the glutelin fraction as a storage protein. As glutelin is rich in all essential amino acids, the nutritional quality of these lines was also supposedly enhanced over the high-yielding parent, Swarna. Mahmoud *et al.* (2008) reported that in a hybrid between IR 64 and *Oryza nivara* the expression of this prolamin band was much higher than that in the two parents. But the nutritional value of prolamins is inferior to glutelins for its low digestibility and negative influence on cooking quality, which increases the hardness of cooked rice (Ogawa *et al.* 1987; Furukawa *et al.* 2003). The breeding lines used by Chattopadhyay *et al.* (2018) showed no significant changes in the intensity of this prolamin band. Moreover, the similar or lower prolamin/glutelin ratio as observed in high-protein lines, compared with Swarna, indicates that the protein quality of the high-protein rice lines remained unchanged if not improved by this breeding programme initiated for quantitative improvement in protein content.

Cypress, an LSU AgCenter-bred semi-dwarf long grain rice known for its excellent grain quality was used to develop and release high protein rice variety, Frontière, in 2017 in USA. It has an averages 10.6% protein which is 54% more than most conventional long-grain rice varieties. Utomo and his team developed a high-protein line of rice cultivar, 'Frontière,' which was released in 2017. The rice was developed through a traditional breeding process. It's the first long grain high-protein rice developed for use anywhere in the world, he says. On average, it has a protein content of 10.6%, a 53% increase from its original protein content. It also needs less heat, time, and usually less water to cook. This high-protein cultivar is currently marketed as "Cahokia" rice. It is grown commercially in Illinois. Hua *et al.,* 2012 reported that a japonica restorer line (named m119) with extremely high protein content was bred successfully, whose protein content (wet base) reached 13.30%±0.19% in brown rice. SDS-PAGE analysis showed that the proportion of glutelin to total protein content increased to 71.77% in m119 compared with the check.

High protein rice through Backcross method of breeding

High protein breeding lines developed through rice pedigree breeding programme at IRRI in 1970s were not probably accepted due to inferior grain quality as compared to popular cultivar IR 8. Backcross method of selection has proved suitable for developing elite introgression lines in the high yielding popular background. This method is also quite effective in producing transgressive segregants for required trait in the background of the high yielding recurrent parent keeping the desired grain quality. Backcross also helps to reduce the effect of undesirable traits (Chattopadhyay *et al.* 2019). Although land races are in general agronomically inferior to high yielding cultivars, transferring favourable alleles from land races to crop plants in grain quality traits has proved achievable in many crops including rice (Yang *et al.*, 2004, Mahmoud *et al.*, 2008; Xie *et al.*, 2014). Chattopadhyay *et al.* (2019a) used one such land race (ARC10075) as high GPC donor in the backcross breeding programme where two to three repeated backcrossing was practiced with the recurrent high yielding parents, cv. Swarna and cv. Naveen for developing two backcross populations (Figure 4). Normal distribution was observed in both the population for GPC with a range of 7.13%-13.6% in Population-1 (ARC10075/Naveen) and 6.09%-14.47% in population-2 (ARC10075/Swarna). BC_3F_3 lines were evaluated for GPC and lines with high GPC (>10%) in the high yielding backgrounds were identified using calibrated NIR spectroscopy. Progenies were screened for higher GPC, grain yield, protein yield and desirable amylose content (20-25%). Through the research carried out by Chattopadhyay *et al.* (2019) it was established that both protein content and grain yield might be improved simultaneously up to a certain extent. This study also reported the qualitative improvement of grain protein in introgressed lines. Except one, all high-protein lines had substantially higher glutelin content than Naveen and Swarna. Glutelin contains mostly essential amino acids and is the chief constituent of protein body II in the endosperm which is better than Protein body I for higher digestibility. Therefore, higher accumulation of glutelin guaranteed better protein quality in those lines. The ratio of prolamin to glutelin fractions ranged from 0.022 to 0.037 which is similar or slightly lower than the adapted parent, Naveen which secured the cooking quality of these introgression lines (Chattopadhyay *et al.,* 2019b). Most of the high-protein lines had noticeably higher levels of essential amino acids such as lysine, threonine, leucine, isoleucine, valine, arginine and also total amino acid compared to recurrent high-yielding parents. Better amino acid profile of some the introgressed lines revealed superior quality of storage protein. This qualitative improvement has been earlier largely limited to maize crop (Ufaz and Galili 2008) through enhancement of grain lysine in QPM lines and has now been extended to rice (Chattopadhyay *et al.,* 2019).

Seven high yielding superior cultures in the background of Naveen (CR2829-PLN-23, CR 2829-PLN-32, CR2829-PLN-37, CR 2829-PLN-98, CR 2829-PLN-99, CR 2829-PLN-100, CR 2829-PLN-114) were nominated for multilocational testing (AICRIP biofortification trial in 2014) (Fig 5). One of those high yielding lines, (CR 2829-PLN-37) with high GPC was released by the Central Variety Re-lease Committee in 2016 for Odisha, Uttar Pradesh and Madhya Pradesh as CR Dhan 310. This was considered as the landmark variety by the Indian Council of Agricultural Research., India (Fig. 6). Another sister line PLN-100 (CR 2829-PLN-100) was released as nutrient-rich rice by State variety Release Committee of Odisha as Mukul (CR Dhan 311) for high protein content (10.1%) and moderately high Zn content (20 ppm), with grain yield of 4330 kg/ha (Fig. 7). Another 3 elite high yielding lines (CR 2830-PLS-17, CR 2830-PLS-124 and CR 2830-PLS-156) with high protein content in the background of cv. Swarna were also evaluated in multilocation at national level (Fig. 8). All these lines were also similar grain type and husk colour to Swarna (Fig. 9). CR 2830-PLS-17 performed at par with Swarna on grain yield basis and found superior than Swarna for grain protein content (10.19%). It had either similar or higher level of essential amino acids such as Lysine (Lys) and Threonine (Thr) as compared to its recurrent parent, Swarna.

Elevated level of some of the essential amino acids improved the quality of the storage protein in this derived line. CR2830-PLS-17 was released as CR Dhan 411 (Swarnanjali) in the state of Odisha (Fig. 10).

The following biofortified rice varieties with elevated level of grain protein content was developed and released by ICAR-NRRI, Cuttack.

a. *CR Dhan 310 (IET24780)*: It has been developed and released in India as the first high protein rice variety for the states of Odisha, Uttar Pradesh and Madhya Pradesh in 2016. It is in the background of high yielding variety cv. Naveen. It has medium duration (120-125 days), semi-dwarf plant type (110 cm) with medium slender and good grain quality. The average grain yield is 4483 kg/ha and it contains average 10.2% protein in polished rice.

b. *Mukul (CR Dhan 311: IET 24772):* It has been released and notified for Odisha as nutrient rich rice in 2019. It is also in the background of cv. Naveen. It has high protein content (10.1%) and moderately high level of Zn content (20 ppm) in 10% polished rice with 5542 kg /ha grain yield in Odisha. It has also medium duration (120-125 days), semi-dwarf plant type (110 cm) with long bold grain and good cooking and eating quality.

c. *CR Dhan 411 (Swarnanjali) (IET 26398: CR Dhan 2830-PLS-17):* This is released in 2021 in Odisha. It is in the background of high yielding variety Swarna (MTU 7029). It had 5621 kg/ha average grain yield. It showed the average protein content and protein yield of 10.01% and 529.2 kg/ha, respectively which were 29% and 31% higher than Swarna.

High protein genotype through inter-species hybridization

Mahmoud *et al.* (2008) found a significant increase in seed protein content in an interspecific hybrid between *Oryza sativa* ssp. *indica* and the wild species *Oryza nivara*. The hybrid showed a protein content of 12.4%, which was 28 and 18.2% higher than those of the parents *O. nivara* and IR 64, respectively. The increase in protein content was dependent on the genetic background of the rice variety used in the hybridization

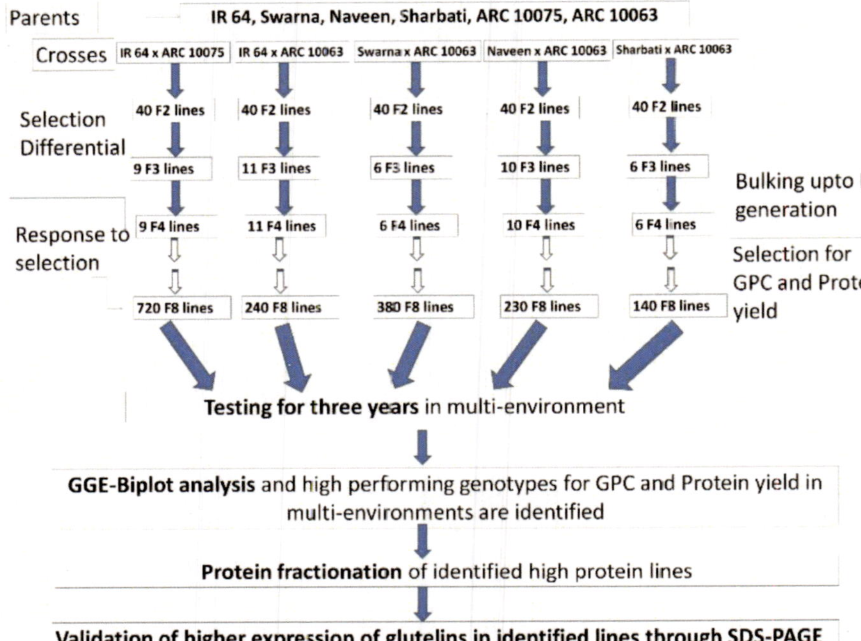

Fig. 2: Schematic diagramme of bulk-pedigree breeding processs followed for qualitative and quantitative improvement of grain protein in rice at ICAR-NRRI, Cuttack (from Chattopadhyay *et al.,* 2018b)

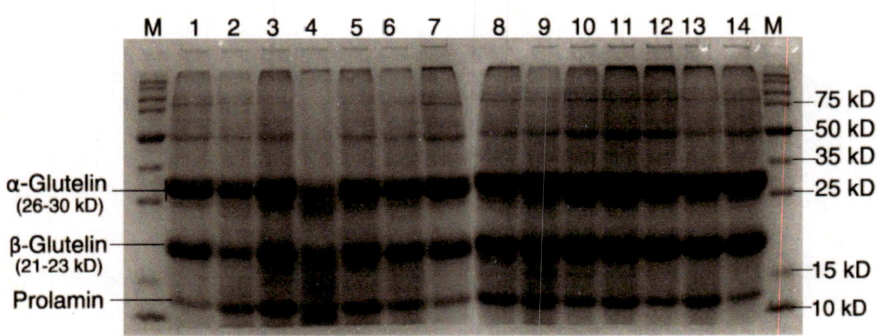

Fig. 3: The SDS-PAGE profiles of partially purified glutelin fraction showing α (~29 kD) and β-glutelin (~21 kD) sub-unit and a prolamin band (~13-14kD) in rice genotypes, viz., lane1: ARC1063, lane 2: Swarna, lane3-14: Breeding lines with high protein and lane M: ladder of 10-225 kD molecular weight. (note: Glutelin 15 μg samples were loaded on each lane and proteins were detected with standard Coomassie brilliant blue stain after electrophoresis)

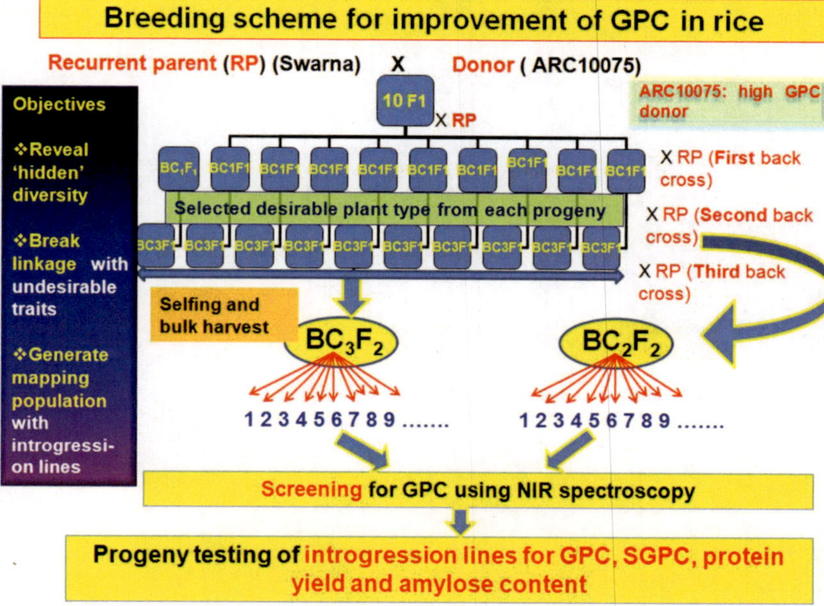

Fig. 4: Breeding scheme for improvement of GPC of high yielding rice through backcross breeding

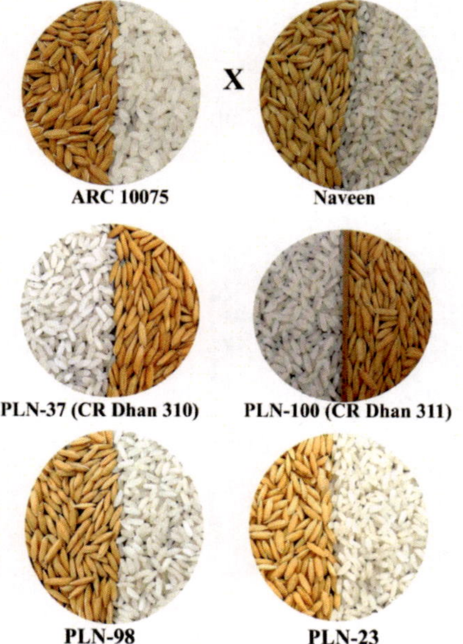

Fig. 5: Grains and milled rice of high protein elite lines in the background of Naveen, derived from Naveen/ ARC 10075 cross

Fig. 6: First high protein landmark rice variety in India, CR Dhan 310

Fig. 7: Biofortified rice varieties CR Dhan 310 and CR Dhan 311 (Mukul), sister lines in the background of cv. Naveen

Fig. 8: CR 2830-PLS-156 (PLS-156) and CR 2830-PLS-17 (PLS-17) introgresssed lines with high protein content in Swarna background along with recurrent parent Swarna and donor parent ARC10075

Fig. 9: Grains, milled rice and panicles of elite high protein lines in the background of Swarna, derived from Swarna/ ARC 10075 cross.

Fig. 10: Field view of Swarnanjali (CR 2830-PLS-17), a high protein Swarna line derived from Swarna/ ARC 10075 cross

References

Aiyswaraya, S., R. Saraswathi, S. Ramchander, R. Vinoth, D. Uma, D. Sudhakar and Robin, S. 2017. An Insight into Total Soluble Proteins across Rice (*Oryza sativa* L.) Germplasm Accessions. *Int.J.Curr.Microbiol.App.Sci*. **6**(12): 2254-2269. doi: https://doi.org/10.20546/*IJCMAS*.612.261

American Society of Agronomy. "High-protein rice brings value, nutrition: Crop breeders build yield, market for high-protein rice." ScienceDaily. ScienceDaily, 23 January 2019. <www.sciencedaily.com/releases/2019/01/190123105809.htm>.

Banerjee, S., Chandel, G., Mandal, N., Meena, B. M., and T. Saluja. 2011. Assessment of nutritive value in milled rice grain of some Indian rice landraces and their molecular characterization. *Bangladesh J. Agril. Res,* **36**(3): 369-380.

Cao, D., M. Hou, Y.S. Guan, M. Jiang, Y. Yang and H.F.Gou. 2009. Expression of HIF1alpha and VEGF in colorectal cancer: association with clinical outcomes and prognostic implications. *BMC Cancer*. **9**: 432

Chandel, G., M. S. Dudhare, T. Saluja, S. M. Shiva, Y. Sharma, A. K. Geda, G. R. Sahu, V. N. Mishra and S. K. Katiyar. 2005. Screening rice accessions for nutritional quality traits to achieve nutritionally balanced rice. In: 5th International Rice Genetics Symposium, Nov. 19-23, 60-61.

Bagchi TB, Sharma S, Chattopadhyay K. 2016. Development of NIRS models to predict protein and amylose content of brown rice and proximate compositions of rice bran. *Food Chemistry*, **191**: 21–27. https://doi.org/10.1016/j.foodchem.2015.05.038

Chattopadhyay K, Das A. and Das S. P. 2011. Grain protein content and genetic diversity of rice in north eastern India. *Oryza* **48**(1):73-75.

Chattopadhyay K, SharmaSG, Bagchi TB, Molla KA, Sarkar S, MarndiBC,Sarkar A, Dash SK, Singh ON. 2018. Development of recombinant high yielding lines with improved protein content in rice (*Oryza sativa* L.). *J Agricul Sci*, 1–17,https://doi.org/10.1017/S0021859618000230

Chattopadhyay K, Behera L, Bagchi TB, Sardar SS, Moharana N, Patra NR, Chakraborti M, Das A, Marndi BC, Sarkar A, Umakanta N, Chakraborty K, Bose LK, Sarkar S, Ray S, Sharma SG. 2019a. Detection of stable QTLs for grain protein content in rice (Oryza sativa L.) employing high throughput phenotyping and genotyping platforms.*Sci Rep*, **9**: 3196,https://doi.org/10.1038/s41598-019-39863-2.

Chattopadhyay K, Sharma S, Bagchi T B, Mohanty B, Sardar S S, Sarkar S and Singh O N (2019a) High-protein rice in high-yielding background, cv. Naveen. *Current Science*. **117** (10) 1722-1726.

da Silveira, C.H., D.E.V. Pires, R.C. Minardi, C. Ribeiro, C.J.M. Veloso, J.C.D. Lopes, W. Meira and G. Neshich G. 2010. Protein cutoff scanning: a comparative analysis of cutoff dependent and cutoff free methods for prospecting contacts in proteins. Proteins: *Struct. Funct. Bioinf.* **74**: 727743

DARE Annual Report 2015–2016, Indian Council of Agricultural Research, New Delhi, 2016. DARE/ICAR Annual Report 2011-12

Frey KJ and Horner T. 1955. Comparison of actual and predicted gains in barley selection experiments. *Agronomy Journal* **47**, 186–188.

FAO 1970.Amino Acid Contents of Food and Biological Data on Protein. FAO Nutritional studies No. 24. Food and Agriculture Organization of the United Nations FAO, Rome.

Furukawa, T., Maekawa, M., Oki, T., Suda, I., Iida, S., Shimada, H., *et al.* (2006). The Rc and Rd genes are involved in proanthocyanidin synthesis in rice pericarp. *Plant J*. **49**, 91–102. doi: 10.1111/j.1365-313X.2006.02958.x

Hua Z T, Ding Y, Wang F, Zhang Q. 2012. Study on protein components and microstructure of the japonica restorer line M119 with high protein content. *Appl Mech Mater*, **140**: 441–445.

ICAR-NRRI Annual Report, 2013-14

ICAR-NRRI Annual Report, 2014-15, https://icar-nrri.in/wp-content/uploads/2018/06/2014-15

ICAR-NRRI Annual report 2019.

Juliano BO. 1993. Rice in Human Nutrition, first ed. International Rice Research Institute, Philippines.

Juliano BO, Bechtel DB. 1985. The rice grain and its gross composition. In: Juliano, B.O.(ed.), Rice Chemistry and Technology, second ed. American Association of Cereal Chemistry, St. Paul, MN, pp. 17-57.

Juliano BO, Villareal CP. 1993. Grain Quality Evaluation of World Rices. Manila: International Rice Research Institute. 205 p.

Juliano BO.2003. Rice: Chemistry and quality. Manila, Philippines. *Phil Rice*: 480.

Juliano BO,Pascaul CG.1980. Quality characteristics of milled rice grown in different countries. IRRI Res. Paper Ser. 48. Int. Rice Res. Inst.: Los Banos, Laguna, Philippines

Juliano, B.O., 1980. Rice: Recent progress in chemistry and nutrition. *In*: Inglett, G. E., Munck, L. (Eds.). Cereals for Food and Beverages, Recent Progress in Cereal Chemistry. Academic Press: New York.

Juliano, B.O. 1985. Criteria and tests for rice grain qualities. In: Juliano BO (ed) Rice chemistry and technology. American Association of Cereal Chemists Inc, *St. Paul* 443–524.

Ju, Z., Hettiarachchy, N. and Rath, N., Extraction, denaturation and hydrophobic properties of rice flour proteins. *J. Food Sci.*, 2001, **66**: 229–232.

Khush, G. S. and Juliano, B. O. 1984. Status of rice varietal improve-ment for protein content at IRRI. In Nuclear Techniques for Cereal Grain Protein Improvement, Proceedings of Research Co-ordination Meeting, International Atomic Energy Agency, Vienna, Austria, pp. 199–202.

Kim JW, *et al.* (2013) Protein content and composition of waxy rice grains. Pakistan Journal of Botany **45**, 151–156.

Kennedy, G and B. Burlingame. 2003. Analysis of food composition data on rice from a plant genetic resources perspective. *Food Chem.* **80**: 589–596.

Krishnan HB and Okita TW.1986. Structural relationship among the rice glutelin polypeptides. Plant Physiology **81**, 748–753.

Kambayashi M, Tsurumi I, Sasahara T. 1984. Genetic Studies on Improvement of Protein Content in Rice Grain. *Japanese Journal of Breeding*, **34(3)**:356–363. https://doi.org/10.1270/jsbbs1951.34.356

Lowry, O. H., Rosebrough, N. J., Lewis Farr, A. and Randall, R. J. 1951. Protein measurement with the Folin phenol reagent. *J. Biol. Chem.*, **193**: 265.

Lee J Y, Kang J W, Jo S M, Kwon Y H, Lee S M, Lee S B, Shin D J, Park D S, Lee J H, Ko J M, Cho J H. 2021. Screening and breeding for biofortification of rice with protein and high lysine contents. Plant Breed Biotechnol, **9(3)**: 199–212.

Liu H, Gan W, Rengel Z, Zhao P (2016) Effects of zinc fertilizer rate and application method on photosynthetic characteristics and grain yield of summer maize. J Soil Sci Plant Nutr **16**:550–562

Mahmoud, A. A., Sukumar, S. and Krishnan, H. B. 2008. Interspecific rice hybrid of *Oryza sativa x Oryza nivara* reveals a significant increase in seed protein content. *J. Agric. Food Chem.*, **52**:476–482.

Mohanty A, Marndi B C, Sharma SG and Das A. 2011. Biochemical characterization of two high protein rice cultivars from Assam rice collections. *Oryza*.**48**(2): 171-174.

Ogawa M, Kumamaru T, Satoh H, Omura T, Iwata N, Omura T, Kasai Z, Tanaka K. 1987. Purification of protein body-I of rice seed and its polypeptide composition. *Plant Cell Physiol* **28**: 1517–1527

Pal P, *et al.* (2016) Effect of nonthermal plasma on physico-chemical, amino acid composition, pasting and protein characteristics of short and long grain rice flour. *Food Research International* **81**, 50–57.

Pal P., Kaur, P., Singh, N., Kaur, A. P., Misra, N. N., Tiwari, B. K., Cullen, P. J., Virdi, A. S., 2016. Effect of nonthermal plasma on physico-chemical, amino acid composition, pasting and protein characteristics of short and long grain rice flour. *Food Research International,* **81**, 50–57.

Patil A.H., Premi V., Sahu V., Dubey M., Sahu GR, Chande G. 2014. Identification of elite rice germplasm lines for grain protein content, SSR based genotyping and DNA fingerprinting. *International Journal of Plant, Animal and Environmental Sciences*, **4**: 127-136

Ren S, Wang S. 2002. Distribution and nutritional analysis of rice proteins. *J Chin Cereals Oils Assoc,* 17: 35–538. (in Chinese with English abstract)

Sompong R, Siebenhandl-Ehn S, Linsberger-Martin G, Berghofer E.2011.Physicochemical and antioxidative properties of red and black ricevarieties from Thailand, China and Sri Lanka, *Food Chem*, **124**: 132–140.

Sambrook J and Russell D (2001) Molecular Cloning: A Laboratory Manual. New York, USA: CSHL Press.

Sambrook J and Russell D (2001) Molecular Cloning: A Laboratory Manual. New York, USA: CSHL Press.

Santos, M.A.; Nicolás, M.F.; Hungria, M. 2006. Identifcation of QTL associated with the symbiosis of Bradyrhizobium japonicum, B. elkanii and soybean = Identifcação de QTL associados à simbiose entre Bradyrhizobium japonicum, B.elkanii e soja. *Pesquisa Agropecuária Brasileira* 41: 67-75 (in Portuguese, with abstract in English).

Shao Y, *et al.* (2011) Infrared spectroscopy and chemometrics for the starch and protein prediction in irradiated rice. Food Chemistry **126**, 1856–1861.

Sugimoto T, Tanaka K and Kasai Z (1986) Improved extraction of rice prolamin. *Agricultural and Biological Chemistry* **50**: 2409–2411.

Tsuzuki, E., Kuroda, H. 1989. Breeding of high protein quality rice. *Euphytica* 43, 47–51 https://doi.org/10.1007/BF00037895

Ufaz S and Galili G. 2008. Improving the content of essential amino acids in crop plants: goals and opportunities. Plant Physiology **147**, 954–961.

Vasal, S.K. 2002. The role of high lysine cereals in animal and human nutrition in Asia. In Protein Sources for the Animal Feed Industry, pp. 167–184. Rome, Italy: FAO.

Wen, T.N. and Luthe, D.S. 1985. Biochemical characterization of rice glutelin. *Plant Physiology* **78**: 172–177.

Wynne, J.C. and Gregory, W.C. 1981. Peanut breeding. *Advances in Agronomy* **34:** 39–71.

Xie, L. H., Tang, S.Q., Chen, N., Luo, J., Jiao, G. A., Shao, G. N., Wei, X.J., Hu, P. S., 2014. Optimisation of near-infrared reflectance model in measuring protein and amylose content of rice flour. *Food Chemistry,* **142**, 92–100.

Xu, Z, Fan, S., Liu, J., Liu, B., Tao, L., Wu, J., Hu, S., Zhao, L., Wang, Q., Wu Y. 2019. A calibration transfer optimized single kernel near-infrared spectroscopic method Spectrochim. *Acta Part A: Mol. Biomol. Spectrosc*., 220, p. 117098

Yang LJ, Xu L, Li JY.2004. Analysis of correlation between protein content, amylose content in the unpolished rice and 1000-grain weight in six different cultivars' rice. *J Shanghai Normal Uni* (Natural Sciences), **10**(Suppl.): 55–58.

Yang Q, Zhang C, Chan M, Zhao D, Chen J, Wang Q, Li Q, Yu H, Gu M,Sun S, Liu Q. 2016. Biofortification of rice with the essential amino acid lysine: Molecular characterization, nutritional evaluation, and field performance. *J Exp Bot* **14**:4285-4296

Yoshida, S., Forno, D. A., Cock, J. H. and Gomez, K. A., 1976. Labora-tory Manual for Physiological Studies of Rice, IRRI, Manila, Phil-ippines, 3rd edn, pp. 1−83.

Yan, W. and Kang, M. S., *GGE Biplot Analysis: A Graphical Tool for Breeders, Geneticists, and Agronomists*, CRC Press, Boca Ra-ton, FL, USA, 2003.

Yu, Y. H., Li, G., Fan, Y.-Y., Zhan, K. Q., Min, J., Zhu, Z. W. and Zhuang, J. Y., 2009. Genetic relationship between grain yield and the contents of protein and fat in a recombinant inbred population of rice. *J. Cereal Sci.*, **50**: 121–125.

Ye, G., Liang, S. and Wan, J., 2010. QTL mapping of protein content in rice using single chromosome segment substation lines. *Theor. Appl. Genet.*, **121**: 741–750.

Zhao L X, Pan T, Cai C H, Wang J, Wei C X. 2016. Application of whole sections of mature cereal seeds to visualize the morphology of endosperm cell and starch and the distribution of storage protein. *J Cereal Sci,* **71**: 19–27.

4

Nutritionally Important Micronutrients (Fe, Zn) and Vitamin (A): Their Biosynthesis, Metabolic Regulation in Rice Molecular Basis of Their Enhancement in Grain

Introduction

Rice is one of the most important staple foods among all cereals. The larger population in Asia and Africa depends on rice for their daily calorie and nutritional requirements. People are found to bank on rice as their main source of nutrition in emerging and underdeveloped nations (Yang *et al*., 2016). However, it is considered that the milled or polished rice is nutritionally poor as the majority of the essential micronutrients *viz*. iron (Fe) and zinc (Zn), and important vitamins are lost during the process of milling and polishing (Johnson *et al*.,2011). So, by the way, the poorer section of the worldwide population mainly depends on rice and is most vulnerable to 'hidden hunger' as they are unable to afford other micronutrients-rich non-staple foods for their balanced diet and are often at the maximum risk for micronutrients deficiencies (Bouis and Saltzman, 2017). The only possible solution for such malnutrition is to have nutritionally enriched food in the daily foodstuff. However, approximately one-third of the world's population is facing the problem of hidden hunger (White and Broadley, 2009). Prolonged utilization of carbohydrate-rich food mainly based on rice, wheat, or maize is contributing to such nutritional deficiency in the poorer section of our society as most of them are unable to supplement nutritionally rich food for their diet. The deficiency of micronutrients can be termed a silent epidemic because it slowly weakens our immune system and hampers physical and intellectual development which may lead to fatality in some severe conditions. In the case of micronutrient deficiencies, deficiencies of iron or iron deficiency anaemia (IDA), zinc deficiency, and vitamin-A deficiency (VAD) are common, and have serious consequences. Globally more than 24,000 people die daily due to "hidden hunger" and malnutrition (Fiyaz *et al.* 2019). Increasing the risk of mortality as well as affecting foetal growth due to malnutrition during pregnancy results in low birth weight (LBW) and risk of survival of the child. The occurrence of malnutrition in developing and underdeveloped countries remains a major public health issue (UNICEF, 2007). To come across this situation, fortification of food with nutritionally enriched micronutrients and essential vitamins could be the probable and continued solution along with alternation of the food processing system. Biofortification of rice or other staple foods could be a game-changer for those people who have very limited choices of dietary resources.

Iron is one of the most important micronutrients among all essential micronutrients, as its deficiency leads to IDA, which is most dangerous to children and women. It can cause cognitive and physical development and weaken our immune system. Even though it can be addressed through nutritional diversification, micronutrient supplements, medicines, etc.; the possibility of treatments may not be accessible to everyone due to geographical and financial incapabilities (Morrissey and Lou, 2009). Likewise, the deficiency of zinc (Zn) is also a significant worldwide public health and nutritional issue (Krishnaswami, 1998; Myers *et al.*, 2014). Less bioavailability of Zn in major cereals, particularly in rice leads to chronic Zn deficiency in people with a rice-based diet. Zn has an important metabolic role in the human body as well as the efficient functioning of cellular metabolic activities as cofactors of 300+ enzymes and stimulation of the immune system (Anderson *et al.*, 2001; Bernett *et al.*, 2010). Although, the outer aleurone layer of hulled or brown rice is the most nutritious form of rice having a relatively higher level of micronutrients like Fe and Zn and important water-soluble vitamins like Vit B1 (Thiamin), B2 (Riboflavin), B3 (Niacin), B5 (Pantothenic acid), B6 (Pyridoxine), B9 (Folate), but lacks other important fat-soluble vitamins like A, D, and K (Ghosh *et al.*, 2019). Therefore, after the processing of hulling and milling a good quantity of these nutritionally important compounds get lost in the white or polished rice. Most importantly, the population of the world that depends on rice tends to prefer white rice over nutritionally superior brown rice for their daily consumption. Therefore, it is very important to improve the bioavailability of these compounds in processed rice i.e. polished white rice.

At this point improving the bioavailability of these essential compounds in food crops through biofortification is of paramount significance. This can be achieved by agronomic management, conventional breeding approaches, or using genetic engineering techniques. Agronomic practices may be performed through soil application of fertilizers or foliar feeding (Phattarakul *et al.* 2012). It may depend upon several key factors like the uptake of nutrients from the soil, its distribution in different plant parts, loss of nutrients during milling or de-husking at the time of food processing, etc. (de Valença *et al.*, 2017). On the contrary, conventional breeding approaches involve identifying and selecting the parent lines containing the desirable traits. Breeding lines are often selected and crossed over for a few generations until the desirable nutritional traits are achieved in the progenies while keeping their agronomic traits intact (Saltzman *et al.* 2017). Furthermore, the advent of modern biotechnological techniques, such as marker-assisted selection has improved significantly the efficiency and precision of breeding processes. More advanced genetic engineering techniques served as an outstanding platform for exploring alternative solutions through genetic modification, which involves removing, altering, or inserting a specific sequence into the plant genome for a specific purpose. It provides greater flexibility by silencing or overexpressing desirable gene sequences for the trait of interest (Abdallah *et al.*, 2015). However, successful biofortification using genetic engineering tools requires in-depth knowledge and understanding of the physiological and molecular processes of uptake, transport, and homeostasis mechanisms in plants.

Iron (Fe) Biofortification in Rice

Iron (Fe) is an important micronutrient for both plants and animals. Fe deficiency is one of the most prevalent micronutrient deficiencies worldwide causing~0.8 million of deaths annually (WHO, 2002). Fe deficiency is ranked in sixth position among the risk factors for death and disability in developing countries with high mortality rates (WHO 2002). Biofortification of rice could be an appropriate approach to improve Fe-deficiency-induced anemia which is a serious health problem in developing countries where rice is the main staple food (WHO 2002; Juliano 1993). Brown rice is rich in mineral value but, rice is mainly consumed in the polished form (the endosperm tissue), which contains low mineral levels (Grusak and Cakmak, 2005). Biofortification can be applied to enhance the Fe concentrations in polished seeds and achieve the target of the Fe requirement for human nutrition.

The iron content in rough rice/ paddy is about 38 ppm. After processing, the iron content in brown rice drops down to 8.8 ppm and ultimately to 4.1 ppm in milled rice (Dexter 1998). As per the reports of Masuda *et al.* (2009), iron content in brown rice was 19 ppm which was reduced by 4.75 times in polished grains with a concentration of 4 ppm. This apparent decline in iron content in the consumable rice grain prompted to initiate the iron biofortification specifically in milled rice. In developing nations, an adequate concentration of iron in rice grain can benefit the health of pregnant women and children. . IDA caused by the shortage of iron has a detrimental effect on human health especially on women and children.Around 32.9% of the people worldwide suffer from IDA, where South Asian and Saharan African nations are at greater risk (Kassebaum *et al.*, 2010). IDA majorly hampers the cognitive development in children, compromises their immune system, and raises the possibility of morbidity. It also have a negative impact on productivity, results in premature birth, and raises the risk of women's mortality. As per the report of WHO's Nutrition Landscape Information System (NLiS), out of 10 densely populated countries, the number of anemic children was reported to be highest (61%) in Pakistan during 2011 and the number of anemic pregnant women was highest (51.5%) in India during 2016. As iron content drops more than any other minerals while post-harvest processing, enhancing iron concentration in the milled rice by adopting biofortification strategies may be beneficial in battling IDA in these affected countries (Dexter, 1998).

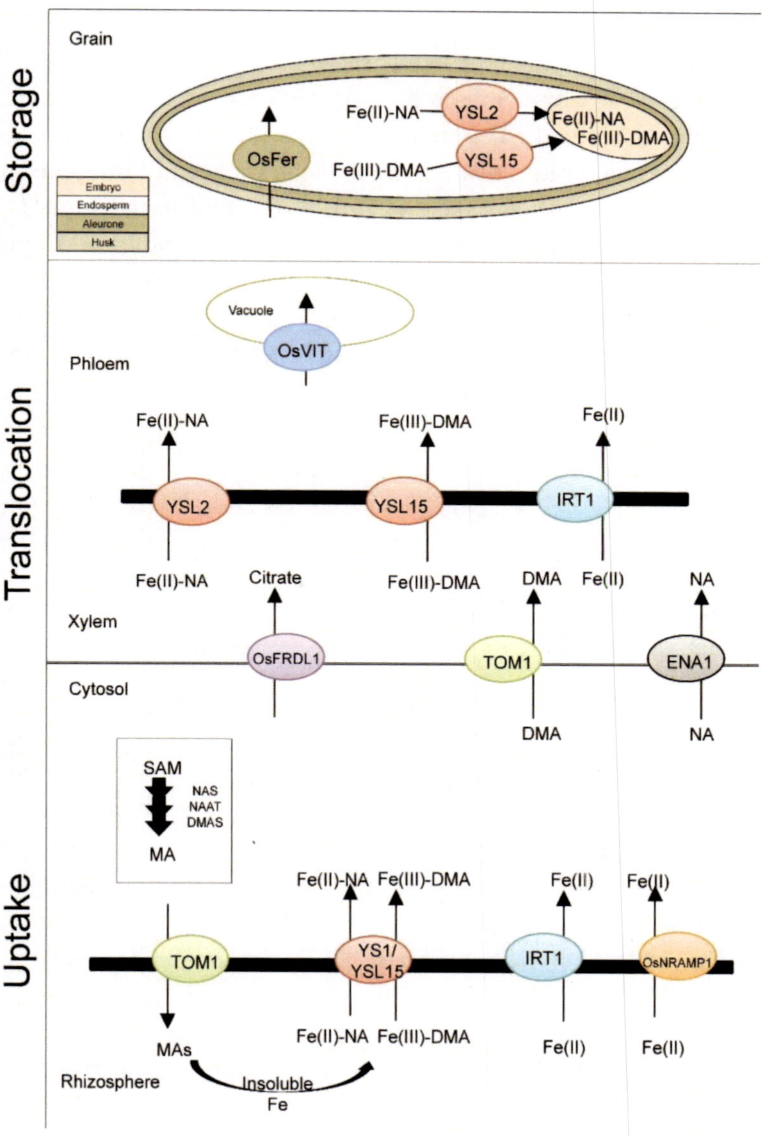

Fig. 1: Schematic representation of iron uptake, transport and storage mechanism in rice. Known genes associated in iron uptake and transport in rice are presented here. TOM1: transporter of mugineic acid family phytosiderophores 1; YS1: yellow strip 1; YSL15: yellow strip 1-like; IRT1: iron-regulated transporter; OsNRAMP1: natural resistance-associated macrophage protein 1; OsFRDL1: ferric reductase defective like1; ENA: efflux transporter of nicotinamide; OsVIT: vacuolar iron transporter; OsFER: ferritin; MA: mugineic acid; DMA: 2'-deoxymugineic acid; NA: nicotinamine; SAM: S-adenosyl-L-methionine; NAS: nicotinamine synthase; NAAT: nicotinamine aminotransferase; DMAS: deooxymugineic acid synthase. The figure is reproduced from Ludwig and Slamet-Loedin (2019) Front. Plant Sci. 10:833.

Iron uptake and its transport in rice

For Fe uptake and accumulation, plants mainly depend upon their rhizospheric sources. There is plenty of Fe available in the soil, but only a minute quantity of Fe is absorbed by the plants. The soil pH and soil redox potential are more important for the availability of Fe content (Morrissey and Lou, 2009). In higher pH, Iron becomes less soluble and it can be found in the form of insoluble ferric oxides. In contrast, it becomes more soluble in acidic pH, from where it can be readily absorbed by the plant roots (Kim and Lou 2007). The uptake of Fe and its distribution in plants are strictly controlled and regulated by different uptake mechanisms, which permit the requisite amount of Fe to be absorbed into the plant while controlling hyperuptake and accumulation, which may exhibit toxicity effects (Welch, 2002). In plants, there are two predominant strategies used for Fe uptake from the soil, *viz.* (i) reduction-based strategy and (ii) chelation-based strategy (Wirth *et al.*, 2009). Rice and other Graminaceous plants predominantly utilize a chelation-based strategy, whereas the reduction-based strategy is more common in non-graminaceous plants. Interestingly, with efficient uptake of Fe in a unique manner, rice can utilize both combinations of reduction-based and chelation-based strategies.

(i) *Reduction-based strategy:* This strategy is mainly utilized by non-graminaceous plants involving the reduction of the ferric (Fe^{3+}) form of available Fe into Fe^{2+} form before being absorbed into the plant system. In the case of the reduction-based strategy, initially, plants release a good amount of protons into the rhizospheric region to decrease the pH of the surroundings under Fe-deficient conditions by the action of different ATPases (Kim and Lou, 2007). The decrease in soil pH thereby increases the solubility of Fe^{3+} in the rhizosphere. In addition, NADPH-dependent Fe chelate reductase also reduces Fe^{3+} into a more soluble Fe^{2+}form with the help of ferric reductase oxidase 2 (FRO2). Finally, this Fe^{2+} form of iron is absorbed and transported into the roots through ferric ion transporters controlled by iron-regulated transporter 1 (IRT1) (Ishimaru *et al.*, 2006).

(ii) *Chelation-based strategy:* Under Fe-deficient conditions, graminaceous plants like rice, wheat, and maize have the unique ability to increase Fe uptake through a chelation-based strategy. Under this strategy, the ferric (Fe^{3+}) form of iron as opposed to the ferrous (Fe^{2+}) form, is directly taken up by the plants from the rhizosphere with the help of soluble phyto-siderophores. Different natural phyto-siderophores like mugineic acid (MA) work as natural iron chelators due to their higher affinity towards the Fe^{3+} form (Morrissey and Lou 2009). Depending on different species, different sets of MAs are released by the plants to the surrounding rhizosphere via the transporter of MAs (TOM1). Rice generally secretes 2′-deoxymugineic acid (DMA) only, while barley secretes different types of MA such as 3-epihydroxymugineic acid (epi-HMA) and 3-epihydroxy-2′-deoxymugineic acid (epi-HDMA) (Ishimaru *et al.* 2006). In case of Fe deficiency, these plants secrete different MAs into the rhizosphere to solubilize the less soluble form of iron (Fe^{3+}) in the rhizosphere. MAs can bind efficiently with Fe^{3+} forming a Fe^{3+}-MA complex, which is usually transported into the root through the yellow stripe 1 (YS1) transporter (Schaaf *et al.*, 2004).

On the other hand, it was reported that rice can utilize both these reduction-based and chelation-based strategies for the uptake of iron in a combined manner (Sperotto *et al.*, 2012). Rice acquires Fe^{3+} via a reduction-based strategy and Fe^{2+} directly from the surroundings through IRT1 or IRT2 transporters. However, a very minimal or no increase in Fe^{3+}-chelate reductase activity was reported in rice roots as found in non-graminaceous plants (Kim and Lou, 2007). As rice predominantly grows in submerged and anaerobic environments rich in Fe^{2+} as compared to the Fe^{3+} form, it could be due to their long-term adaptation to these environments (Sperotto *et al.*, 2012). But, like other graminaceous plants, rice also secrets MAs into the rhizosphere to bind with Fe^{3+} and make Fe^{3+}-MA complexes, which are transported into the root via YS-like 15 (YSL15). Among both strategies, rice is capable of iron uptake from the surroundings more efficiently through Fe^{3+}-MA complexes as compared to direct Fe^{2+} uptake (Ishimaru *et al.*, 2006).

Enhancement of Iron Availability in Plants

As discussed above, the uptake of iron and its transportation and accumulation into the plant is highly regulated. Thus, the bioavailability of Fe can be raised by agronomic interventions such as improved fertilizer administration to the soil or by directly foliar feeding on to the leaf surface. Direct foliar application elevates the efficiency of iron uptake and transport in rice plants which was comparatively lesser in soil fertilization (Yuan *et al.*, 2013). Nevertheless, rain frequently washes away the foliar applied fertilizers, necessitating reapplication after each precipitation, which is costly and may also have severe effects on the environment (García-bañuelos ML, Sida-arreola, 2014). As against that increasing the Fe content in rice seems to be a much safer and more sustainable option for Fe biofortification.

Pro-Vitamin A biofortification in Rice:

Apart from a few micronutrients, rice grain is also deficient in provitamin-A (Yo *et al.*, 2000). Vitamin-A is generally produced from beta-carotene and plays a pivotal role in maintaining physiological functions in the animal body like vision, growth, reproduction, cellular differentiation, and immunity. It is also essential for adult gene regulation. Animals cannot produce carotenoids in their body and for this, they are dependent on plants. The majority of cultivated rice grown in lands are produced beta-carotene in their leaves and stems, but not in the endosperm. Molecular analysis and detailed studies in this regard have identified that the process of beta-carotene synthesis is blocked in rice endosperm. Therefore, the predominant consumption of rice for fulfilling energy and nutritional demands often promotes vitamin-A deficiency in animal bodies (Yo *et al.*, 2000). Vitamin-A deficiency is important due to its association with night blindness and body immunity (Tang *et al.* 2009; Farré *et al.*, 2011). Based on the reports, worldwide, 250 million preschool children become blind every year, due to the deficiency of vitamin-A, and 10% of these are due to increased susceptibility to infectious diseases (Krishnan *et al.*, 2009; Brown and Noelle, 2015). Therefore, the development of a rice genotype, which meets the Vitamin-A demand of those poor people is considered a prime need for rice grain improvement programmes and could be a hope to combat Vitamin-A deficiency.

To combat this issue and to alleviate the bioavailability of vitamin-A in rice endosperm, initially, Ingo Potrykur and Peter Beyer started their work and developed rice with golden endosperm, which is popularly known as 'Golden rice'. Golden rice is named due to its yellowish colour of kernels or polished grain. This novel variety can accumulate provitamin-A in the endosperm. Later, other scientists from IRRI (International Rice Research Institute) and different parts of the world also performed this process to improve the overall carotenoid content of rice endosperm. After its development, Golden rice has gone through several obstructions and difficulties for its field release as it is developed employing the transgenic approach. Therefore, bio-safety measures and the nutritional assessment process were much more prolonged here than in conventional breeding. Based on the most recent reports, BRRI (Bangladesh Rice Research Institute) developed 'BRRI Dhan 29' for the production of carotenoids in rice endosperm, also very close to releasing this genotype for open field cultivation and production of carotenoid-enriched rice grains (Amna *et al.* 2020).

Biosynthesis Pathway of Beta-Carotene in Plants and Non-Photosynthetic Bacteria and Fungi:

Carotenoids have widely distributed a class of pigments, which contains 40 carbons, generally present in all photosynthetic organisms and also found in some non-photosynthetic organisms like bacteria and fungi. The biosynthesis of carotenoids takes place in plastids; in the case of photosynthetic cells it takes place in chloroplasts and for fruits, it generally takes place in chromoplasts. In plants, biosynthesis of carotenoids is a four-step process, which starts with a compound GGPP, a 20-carbon compound (geranyl geranyl diphosphate), ultimately 2- molecules of GGPP converted to produce a different carotenoid compound, which is 40 carbons. (Britton, 1988; Cunningham and Gantt, 1998; Sandmann, 1994, 2001). In general, carotenoids, chlorophylls, and other compounds are produced from the same biosynthetic precursor compound GGPP. This is produced from the IPP/ DMAPP (C5) compound which is the isopentenyl-diphosphate (IPP) pathway along with its isomer DMAPP (dimethylallyl diphosphate by the activity of different enzymes. Then GGPP was then converted into phytoene by the activity of phytoene synthase (PSY). Finally, from phytoene in plants, the activity of three different enzymes phytoene desaturase (PDS), zeta-carotene desaturase (ZDS), and lycopene cyclase further, complete the pathway of beta-carotene production/ synthesis. In rice leaves and stems, this cycle was present; unfortunately, this process was somehow blocked in the rice endosperm. In plants, PDS and ZDS catalyze the conversion of phytoene to lycopene, and then lycopene is converted into beta-carotene by the activity of the lycopene cyclase enzyme. This process was slightly different in the case of non-photosynthetic bacteria and fungi. Though the precursor compound was the same in plants and non-photosynthetic organisms and it is GGPP, the bacterial cell does not possess PDS and ZDS, instead of the function of two different enzymes (PDS and ZDS) in bacterial cell, functions of two different enzymes are stably maintained by the activity of a single enzyme (CRTI), which catalyzes the conversion of phytoene to lycopene and further helps in the biosynthesis of beta-carotene.

Zinc (Zn) Bio-fortification in Rice

Zinc plays an important role in the regulation and iron absorption in the intestine and an adequate amount of zinc (along with iron) is essential for IDA treatment (Graham *et al*. 2012). Zinc is also essential for growth and development, immune system functioning, and reproductive, sensory, and neurobehavioral development. For the activation of more than 300 enzymes and proteins, zinc is the main essential component, as zinc is the only metal that is involved in the structure and function of all six classes of enzymes (Levenson *et al*. 2011; Broadley *et al*. 2007; Maret 2013).

Zinc is required for the proper function of different transcription factors and zinc finger proteins. There are about 17.3% of the world's population is affected by zinc deficiency and under the age of five more than 400 million children die every year due to zinc deficiency (Hefferon 2019). Deficiency of Zn affects severe difficulties of health issues in children like diarrhea, weight loss, growth inhibition, and anorexia. Changes in infants' neurobehavioral are noticeable while changes in skin and dwarfing are more common in toddlers and school-going children (Prasad *et al.*, 2014).

As Zinc is an essential micronutrient, it plays an important role in plants. It acts as a cofactor and activators of various types of hormones e.g. auxin which is necessary for growth and development in the plant (Begum *et al.*, 2016). There are different biochemical processes like the production of nucleotides, auxin metabolism, activation of the enzyme, and formation of chlorophyll that are regulated by Zn and also play an important role in fertilization as a high amount of Zn is present in pollen grains. Zinc also plays an important role in physiological growth and development including the synthesis of protein, carbohydrates, lipids, and nucleic acids, expression, and regulation of a gene in the plant (Chang *et al.*, 2005). For the maintenance of the functional and structural integrity of biological membranes, zinc is required mostly due to its binding to SH-containing compounds (Sadeghzadeh & Rengel, 2011). Zn is an integral part of Cu/Zn-super-oxide dismutase (SOD) and also plays a key role in the detoxification of reactive oxygen species (ROS) (Cakmak 2000).

A limited nutritional range and insufficiency of crucial minerals like zinc (Zn) distinguish the diets of malnourished people. Deficiency of Zn is a worldwide dietary difficulty and a major issue in developing countries. Cereal grains are the solution to fulfill human daily dietary needs, but they have a very minimal quantity of Zn in the grain, particularly when grown in soil having insufficient Zn concentration. Zinc insufficiency can be preserved in numerous ways viz., diversification of nutrition, enrichment of food, and biofortification. There are several restrictions concerning the nutritional diversification and food enrichment preferred biofortification of Zn as a continuous solution to undernourishment. Among the possible bio-fortification choices to resolve Zn deficiency, plant breeding methods, and agronomic biofortification suggest the most important advantage.

Zn Uptake, distribution, and accumulation in plants

It was widely reported that the rice grown in lowlands is continuously under the submerged environment where low availability of Zn is prevalent (Fageria 2013; Meng *et al.*, 2014). Zinc present in the soil is transported to the root by mass flow and the concentration is augmented by diffusion and root extension

(McGrath and Lobell 2013; Yoneyama *et al.* 2015). Though a minute quantity of Zinc may have been absorbed by the root and transferred to the xylem through an apoplastic pathway, mainly the symplastic pathway plays an important role in the transport of Zn across the roots to the xylem (Broadley *et al.*, 2007). The uptake of Zn by active transport and the energy demand is a light reaction of photosynthesis (Behrenfeld *et al.* 2004; Yoneyama *et al.*, 2015).

During the symplastic movement of Zinc, it enters the apoplast before it is taken into the new symplast (Olsen and Palmgren, 2014). Generally, Zn may be found in as Zn–phytosiderophore complex or in the form of Zn^{2+} ions (Broadley *et al.*, 2007). There are several Ca^{2+} channels to mediate Zn^{2+} which are present in the plasma membrane. But, Zn is mostly regulated by ZIPs e.g. ZIP1, ZIP3, and ZIP4 (Humayan Kabir *et al.* 2014; Bashir *et al.* 2012, Palmgren *et al.* 2008). MTPs (Arrivault *et al.* 2006), HMAs (Papoyan & Kochian 2004), YSLs (Suzuki *et al.*, 2006), NASs (Talke *et al.* 2006), (Haydon & Cobbett 2007), ZIPs (Milner *et al.*, 2013), and ZIF1 are the most essential genes which are involved in phytosiderophores biosynthesis. There are lots of Zn^{2+} holding proteins in plant cytoplasm, but the concentration of Zn^{2+} happens to be very low in general (Broadley *et al.*, 2007). Zn may travel within the xylem as Zn^{2+} or in the form of a complex with organic acids, nicotinamide, or histidine. On the other hand, Zn is considered to be stored in the vacuole as complexes of organic acids (Leitenmaier & Küpper, 2013). By the members of the ZIP family, Zn^{2+} influx is mediated to a portion of the leaf and ultimately to the phloem (Ishimaru *et al.* 2005). Furthermore, Zn transport by YSL proteins in the phloem may transfer Zn as a form of a complex with small proteins.

Zinc allocation between plant parts plays a significant role in determining the concentration of Zn in the grain. Zinc distribution within various plant parts is subjected to the stage of physiological growth and the presence of nutrients in the plant (Seneweera 2011; Impa *et al.* 2013; McGrath and Lobell 2013; Shivay *et al.*, 2015; Yoneyama *et al.*, 2015). Depending on the requirement of tissue, first Zn is distributed to leaves during vegetative growth and to grains during grain filling, which is extremely metabolic active sinks, and afterward to the stem and sheath (Jiang *et al.*, 2008). At the later stage of development, the allocation of Zn between the grain, leaf blade, and shoots is also different (Seneweera, 2011). At lower Zinc concentrations, the grain yield decreases so it is very significant to keep at least the minimum internal Zn concentration. Thus, the concentration of Zn in grain may be determined by the transport of Zn from the roots, stems, and leaves (Stomph *et al.*, 2009; Jiang *et al.*, 2007).

Transporters and transcription factors for Zn Uptake

There are some transporters like natural resistance-associated macrophage protein (NRAMP) and Fe-regulated transporter (IRT) which are associated with iron transport, while for Zinc transport only ZIPS family transporters are known to play a role. These transporters were isolated from several crop plants and have shown diverse functions. It was reported that ZRT/IRT-like proteins and ZIP-like transporters e.g. AtZIP1, AtZIP2, AtZIP3, and AtZIP4 were required for the uptake of Zn into the roots (Guerinot, 2000). Transport of mineral elements of the roots to the xylem, where they are transported to other plant parts. Zn was probably chelated by nicotinamide (NA) both in the xylem and phloem during its transport (Von Wirén *et al.*, 1999). Zinc is present in the xylem bound to small proteins or phytate and its transportation is mediated by the transpirational pool (Sadeghzadeh, 2013). Zinc present in the phloem is bound with nicotinamide which is the main ligand in rice phloem sap (Nishiyama *et al.*, 2012). There is a minute amount of iron/zinc that remains in roots as stock and is used when required for growth from the above-ground parts. NRAMP protein transporters play an important role in the vacuolar pool and remobilization of minerals (Thomine *et al.*, 2003). Nakandalage *et al.* (2016) reported the mass flow of Zn uptake and transport for loading into the rice grain. The transporter genes of the ZIP family such as *OsZIP4, OsZIP5,* and *OsZIP8* are associated with Zn transport from root to shoot. The unloading of Zn from the xylem and allocation of Zn to the developing tissues is regulated by OsZIP3 (Sasaki *et al.*, 2015). Although the involvement of ZIP in the homeostasis of Zn, *OsHMA2* (P-type heavy metal ATPase) homolog also plays a role in root-to-shoot zinc transport as well as implicated in a better supply of Zn to the developing tissue in rice (Sasaki *et al.*, 2015; Satoh- Nagasawa *et al.* 2012; Takahashi *et al.*, 2012).

Ferritin protein plays an important role in the accumulation of Fe and Zn in the seed. By transferring the soybean ferritin genes in rice, Fe and Zn content has been increased in seeds (Vasconcelos *et al.*, 2003). The gene of ferritin was transferred either from the French bean or soybean into rice (Lucca *et al.* 2001; Qu *et al.*, 2005) which improved the grain Fe and Zn contents. NAC transcription factor (NAM-B1) which is encoded by the Gpc-B1 locus of *Triticum diccocoides*, plays a role in increasing the zinc and iron contents in the grain. Downregulation of NAM genes might delay senescence and reduction of Zn and Fe content. The up-regulation of the NAM genes (Uauy *et al.*, 2006), ferritin (Drakakaki *et al.*, 2005), and Zn transporter protein (Ramesh *et al.* 2004) have not revealed any relation with agronomic traits.

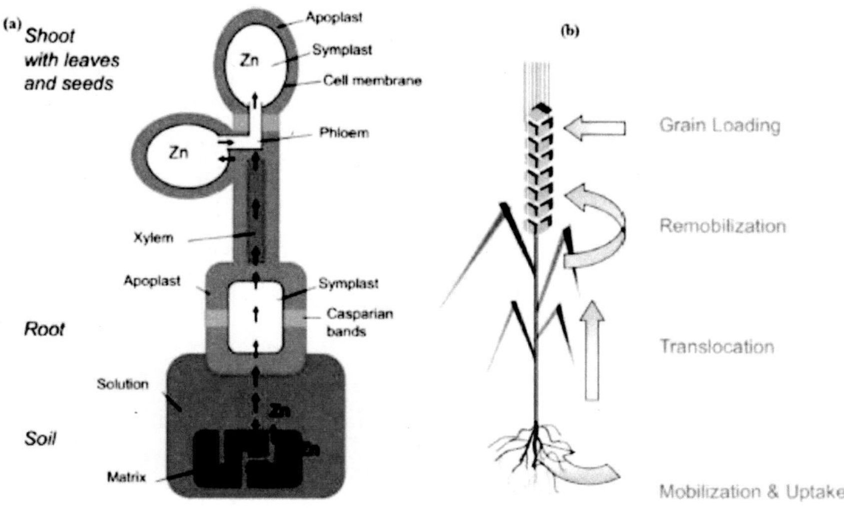

Fig. 2:(a) Zn transfer from soil into grains: 1-Zn release from soil matrix into solution through dissolution of solid phases and desorption from partide surfaces, 2-movement to roots via diffusion, mass flow with water or transport in hyphae of mycorrhizal fungi, 3-uptake into root cell symplast and transfer into xylem (both of which involves passage of cell membranes), 4-translocation through xylem from roots into shoots with transpiration stream and allocation in leaf cells, 5 relocation via phloem (=symplasmic transport) from leaves into grains during grain development (adopted from Schulin *et al.*, 2015). (b) Increasing grain Zn density through organic matter inputs into soil: 1-mobilization and uptake, 2-translocation, 3-remobilization, 4-grain loading (adopted from Cakmak *et al.*, 2010).

Enhancement of Zn Availability in Plants

As discussed above, Zn uptake and its distribution and accumulation into plant parts are highly synchronized. To improve the availability of plants, access, and employment with high content of important nutrients, some changes are necessary which may include; the provision of improved cereal varieties, adding Zn enhancer for absorption, improvement of customized milling practices, increasing the phytase activity of microbes, and bio-fortification of crops with Zn (De Valença *et al.*, 2017). The two probable approaches for Zn biofortification are – (i) agronomic and (ii) genetic biofortification.

(i) *Agronomic biofortification:* Enhancement of bioavailability of Zn can be governed by agronomic manipulations *viz.* bio-fortification by fertilizer application is a direct method to enhance the mineral content and yield of grain on infertile soils (Mao *et al.*, 2014). Zn fertilizers application in the soil is a common approach to manage deficiencies of Zn and to enhance Zn concentration in grain. Foliar application is better effective in grain Zn build-up than soil application (Jan *et al.*, 2016). On the contrary, soil application is more effective in increasing grain yields (Cakmak *et al.*, 2010; Liu *et al.*, 2016). The application of both combined foliar + soil application is a superior approach for both grain Zn uptake and grain yield (Gomez-Coronado *et al.* 2016). When Zinc was applied to rice through soil treatment, the maximum Zn biofortification was observed (Nattinee *et al.*, 2009). In addition, better uptake of Zn into the leaves was found when Zinc was applied with urea fertilizer.

Treatment of seeds with a high concentration of Zn at the sowing stage and application of Zn fertilizer at particular stages to increase both grain Zn content and grain yield (Stomph *et al.* 2011). Treatment of seed with Zn can be categorized in two ways: seed coating and seed priming (Farooq *et al.*, 2012). The priming of seeds is an easy and low-cost method of soaking seeds in solutions of essential nutrients for a certain time. As compared to soil application, seed treatment is a comparatively better choice due to the lesser amount of nutrient requirement. This helps in improved germination and growth of seedlings under different abiotic stresses.

(ii) *Genetic biofortification:* By genetic biofortification, new crop varieties can be developed that possess comparatively better Zn accumulating capability in the grains. Genetic biofortification through conventional breeding approaches by selecting better-quality genotypes was used to decrease the prevailing Zn deficiency in rice through a more reasonable, cost-effective approach appropriate to the desired populations (Velu *et al.*, 2014).

It was found that differences in genetic concentrations of Zn in the grains of diverse germplasms of rice and maize. Such genetic dissimilarity is usually absent in the modern-day cultivated varieties developed through breeding programmes. On the other hand, it is considered that the application of molecular physiology and genetic control are the major sources of high Zn content and other micronutrients in cereal grains (Lucca *et al.*, 2006; Johnson *et al.*, 2011). Several studies have reported that the genes which are responsible for higher Zn content in grains, are most probable to carry out the specific genes (Cakmak *et al.*, 2004). Locus Gpc-B1 which influences both protein and Zn concentrations is present on chromosome 6B (Distelfeld *et al.*, 2007). Distelfeld *et al.* (2007) showed a very strong relationship between protein and Zinc concentrations in rice grains. The genes responsible for high grain Zn content were also found to regulate the grain protein content due to a close positive relationship between these two traits. Genetic factors that are responsible for protein content and Zn concentration are possibly co-segregated in cereals (Morgounov *et al.* 2007).

In Asian countries such as Bangladesh, rice is the major source of zinc intake, where rice alone provides 69% of nutritional zinc to women and 49% of zinc to children (Arsenault *et al.*, 2010). CGIAR-Harvest Plus released a zinc biofortified rice variety in 2013 which was developed by conventional breeding in Bangladesh. Recently about 1.5 million households farming established zinc-biofortified rice of eight varieties and have since been growing them (Goldstein 2018). A medium duration (125 days) biofortified semi-dwarf variety with a non-lodging plant variety named IET 23832 (DRR Dhan 45) with 22.6–24.0 ppm zinc concentration in polished grain developed by IIRR, Hyderabad in 2019. Through conventional breeding, IET 23832 was developed by using Harvest Plus material with several essential qualities like desirable amylose content (20.7%), ensuring the quality of good cooking and moderate resistance to sheath rot disease (*Sarocladium oryzae*), rice blast disease (*Magnaporthe grisea*) and rice tungro virus infection.

As we have discussed above the functions of Zinc in rice, uptake, and allocation of Zn, and the comparative efficacies of diversified sources of Zn applied by modern and conventional approaches to increase soil and bioavailability of Zinc in rice. Moreover, we have discussed agronomic management and breeding methods to increase Zinc uptake and distribution into rice grains for the enhancement of quality and yield that will eventually help to ensure food and health security for developing countries.

References

Abdallah NA, Prakash CS, McHughen AG. 2015. Genome editing for crop improvement: Challenges and opportunities. *GM Crops Food.* **6**:183-205.

Ali N, Datta SK, Datt K. 2010. RNA interference in designing transgenic crops. GM Crops 1:207–213.

Ali N, Paul S, Gayen D, Sarkar SN, Datta SK, Datta K 2013. RNAi mediated down regulation of myo-inositol-3-phosphate synthase to generate low phytate Rice **6**:12. doi: 10.1186/1939-8433-6-12.

Amna SQ, Tantray AY, Bashir SS, Zaid A, Wani SH. 2020. Golden Rice: Genetic Engineering, Promises, Present Status and Future Prospects. In: Rice Research for quality improvement: Genomics and Genetic engineering (Eds. Aryadeep Roychoudhury) pp: 581-603.

Anderson RA, Roussel AM, Zouari N, Mahjoub S, Matheau JM, Kerkeni A. 2001. Potential antioxidant effects of zinc and chromium supplementation in people with type 2 diabetes mellitus. *J Am Coll Nutr* 20:212-8.

Arrivault S, Senger T, Krämer U 2006. The Arabidopsis metal tolerance protein AtMTP3 maintains metal homeostasis by mediating Zn exclusion from the shoot under Fe deficiency and Zn over supply. *Plant J.* **46**:861–879.

Arsenault JE, Yakes EA, Hossain, MB, Islam MM, Ahmed T, Hotz C, Lewis B, Rahman AS, Jamil KM, Brown KH. 2010. The current high prevalence of dietary zinc inadequacy among children and womenin rural Bangladesh could be substantially ameliorated by zinc biofortification of rice. *J Nutr* **140**:1683–1690

Barnett SA, Amyes TL, Wood BM, McKay B, Gerlt JA Richard JP. 2010. Activation of R235A Mutant Orotidine 5′-Monophosphate Decarboxylase by the Guanidinium Cation: Effective Molarity of the Cationic Side Chain of Arg-235. *Biochemistry,* **49:**824–826

Bashir K, Ishimaru Y, Nishizawa NK 2012. Molecular mechanisms of zinc uptake and translocation in rice. *Plant Soil*, **361**:189–201

Bashir K, Takahashi R, Akhtar S, Ishimaru Y, Nakanishi H, Nishizawa NK. 2013. The knockdown of OsVIT2 and MIT affects iron localization in rice seed. *Rice* **6**(1):31

Begum MC, Islam M, Sarkar MR, Azad MAS, Huda AN, Kabir AH 2016. Auxin signaling is closely associated with Znefficiency in rice (*Oryza sativa* L.). *J Plant Interact*, **11**:124–129

Behrenfeld MJ, Prasil O, Babin M, Bruyant F 2004. In search of a physiological basis for covariations in light limited and light saturated photosynthesis. *J. Phycol.,* **40**:4–25

Boonyaves K, Wu TY, Gruissem W, Bhullar NK. 2017 Enhanced Grain Iron Levels in Rice Expressing an IRON-REGULATED METAL TRANSPORTER, NICOTIANAMINE SYNTHASE, and FERRITIN Gene Cassette. *Front Plant Sci* **8**: 130

Borrill P, Connorton JM, Balk J, Miller AJ, Sanders D, Uauy C. 2014. Biofortification of wheat grain with iron and zinc: integrating novel genomic resources and knowledge from model crops. Front Plant Sci 21:53–60

Bouis HE, Saltzman A (2017) Improving nutrition through biofortification: A review of evidence from HarvestPlus, 2003 through 2016. Vol. 12, Global Food Security. Elsevier; p. 49-58

Brinch-Pedersen H, Borg S, Tauris B, Holm PB. 2007. Molecular genetic approaches to increasing mineral availability and vitamin content of cereals. *J Cereal Sci* 46:308–326

Britton G. 1988. Biosynthesis of carotenoids. In: Plant Pigments (TW. Godwin, ed.), pp. 133–18

Broadley MR, White PJ, Hammond JP, Zelko I, Lux A. 2007. Zn in plants. New Phytol 173:677–702

Brown C.C., Noelle R.J. 2015. Seeing through the dark: new insights into the immune regulatory functions of vitamin A. *Eur. J. Immunol.* **45:** 1287–1295.

Cakmak I 2000. Role of Zn in protecting plant cells from reactive oxygen species. *New Phytol* **146**:185–205.

Cakmak I, Pfeiffer WH, McClafferty B. 2010. Review: biofortification of durum wheat with zinc and iron. *Cereal Chem* **87:**10–20.

Cakmak I, Torun A, Millet E, Feldman M, Fahima T, Korol A, Nevo E, Braun HJ, Özkan H 2004. Triticum dicoccoides: an important genetic resource for increasing zinc and iron concentration in modern cultivated wheat. *Soil Sci Plant Nutr* **50:**1047–1054.

Chang HB, Lin CW, Huang HJ 2005. Zinc-induced cell death in rice (*Oryza sativa* L.) roots. *Plant Growth Regul* **46**:261–266.

Cunningham FX. and Gantt E 1998. Genes and enzymes in carotenoid biosynthsis. Annu. Rev. Physiol. *Plant Mol. Biol*. **49:** 557– 583

Datta K, Baisakh N, Oliva N, Torrizo L, Abrigo E, Tan J, *et al*. 2003. Bioengineered ‘golden’ indica rice cultivars with β-carotene metabolism in the endosperm with hygromycin and mannose selection systems. *Plant Biotechnol. J.*, **1:** 81–90.

De Valença AW, Bake A, Brouwer ID, Giller KE. 2017. Agronomic biofortification of crops to fight hidden hunger in sub- Saharan Africa. *Glob Food Sec.,* **12:**8–14

Dexter, P.B. 1998. Rice Fortification for Developing Countries; OMNI/USAID: Washington, DC, USA

Distelfeld, A., Cakmak, I., Peleg, Z., Ozturk, L., Yazici, A.M., Budak, H., Saranga, Y., Fahima T. 2007. Multiple QTL-effects of wheat Gpc-B1 locus on grain protein and micronutrient concentrations. *Plant Physiol* **129**:635–643

Drakakaki, G., Marcel, S., Glahn, R.P., Lund, E.K., Pariagh, S., Fischer, R., Christou, P., Stoger E. 2005. Endosperm specific coexpression of recombinant soybean ferritin and Aspergillus phytase in maize results in significant increases in the levels of bioavailable iron. *Plant Mol Biol* **59:**869–880

Fageria, N.K. 2013. Mineral Nutrition of Rice. Boca Raton, FL: CRC Press.doi: 10.1201/b15392

Farooq M, Wahid A, Siddique KHM 2012. Micronutrient application through seed treatments: a review. J *Soil Sci Plant Nutr* **12:**125–142.

Farré G, Maiam Rivera S, Alves R, Vilaprinyo E, Sorribas A, Canela R *et al.* 2013. Targeted transcriptomic and metabolic profiling reveals temporal bottlenecks in the maize carotenoid pathway that may be addressed by multigene engineering. *Plant J.* **75**:441–455.

Fiaz S, Ahmad S, Noor MA, Wang X, Younas A, Riaz A, Riaz A, Ali F 2019. Applications of the CRISPR/Cas9 system for rice grain quality improvement: Perspectives and opportunities. *Int J Mol Sci.*, **20**:888

García-bañuelos ML, Sida-arreola JP, Sanchez E 2014. Biofortification- Promising approach to increasing the content of iron and zinc in staple food crops. J Elem 19:865-888

Gayen D, Ghosh S, Paul S, Sarkar SN, Datta SK, Datta K 2016. Metabolic Regulation of Carotenoid-Enriched Golden Rice Line. *Front Plant Sci.* **7:**1622.

Ghosh S, Datta K, Datta SK. 2019. Rice vitamins. In Rice Chemistry and Technology, 1st ed.; Bao, J., Ed.; Elsevier Inc.: Amsterdam, The Netherlands; AACC International: St. Paul, MN, USA, **7:**195–220.

Goldstein P 2018. Harvest Plus Talks Zinc Rice with Farmers in South Eastern Bangladesh.

Gomez-Coronado F, Poblaciones MJ, Almeida AS, Cakmak I. 2016. Zinc (Zn) concentration of bread wheat grown under Mediterranean conditions as affected by genotype and soil/foliar Zn application. *Plant Soil.* **401**:331–346.

Graham RD, Knez M, Welch RM 2012. How much nutritional iron deficiency in humans globally is due to an underlying zinc deficiency? *Adv Agron.* **115**:1–40

Grusak MA, Cakmak I 2005. Methods to improve the crop delivery of minerals to humans and livestock. In: Broadley MR, White PJ (eds) Plant Nutritional Genomics. Blackwell, Oxford, pp 265–286.

Haydon MJ, Cobbett CS 2007. A novel major facilitator superfamily protein at the tonoplast influences zinc tolerance and accumulation in Arabidopsis. Plant Physiol 143:1705–1719

Hefferon K (2019) Biotechnological approaches for generating zinc-enriched crops to combat malnutrition. *Nutrients* **11:** 253.

Humayan Kabir A, Swaraz AM Stangoulis J. 2014. Zinc-deficiency resistance and biofortification in plants. *J Plant Nutr Soil Sci.,***177**:311–319.

Impa SM, Gramlich A, Tandy S, Schulin R, Frossard E, Johnson Beebout SE (2013) Internal Zn allocation influences Zn deficiency tolerance and grain Zn loading in rice (*Oryzasativa* L.). *Front Plant Sc.*, **4:**534.

Ishimaru Y, Masuda H, Bashir K, Inoue H, Tsukamoto T, Takahashi M, Nakanish H, Aoki N, Hirose T, Ohsugi R, Nishizawa NK. 2010. Rice Metalnicotianamine Transporter, OsYSL2, is required for the long-distance transport of iron and manganese. *Plant J.* **62:**379-390

Ishimaru Y, Masuda H, Suzuki M, Bashir K, Takahashi M, NakanishiH, Mori S, Nishizawa NK 2007. Overexpression of the OsZIP4 zinc transporter confers disarrangement of zinc distribution in rice plants. *J Exp Bot* **58:**2909–2915

Ishimaru Y, Suzuki M, Kobayashi T, Takahashi M, Nakanishi H, Mori S, Nishizawa NK. 2005. OsZIP4, a novel Zn-regulated Zn transporter in rice. *J Exp Bot.*, **56:**3207–3214

Ishimaru Y, Suzuki M, Tsukamoto T, Suzuki K, Nakazono M, Kobayashi T, Wada Y, Watanabe S, Matsuhashi S, Takahashi M, Nakanishi H, Mori S, Nishizawa NK 2006. Rice plants take up iron as an Fe3+-phytosiderophore and as Fe2+. *Plant J.* **45:**335-346

Jan M, Anwar-ul-Haq M, Tanveer-ul-Haq AA, Wariach EA 2016. Evaluation of soil and foliar applied Zn sources on rice (*Oryza sativa* L.) genotypes in saline environments. *Int J Agricul Biol.* **18:**643–648.

Jiang W, Struik PC, Lingna J, VanKeulen H, Ming Z, Stomph TJ. 2007. Uptake and distribution of root-applied or foliar-applied Zn after flowering in aerobic rice. *Ann Appl Biol.* **150**: 383–391.

Jiang W, Struik PC, VanKeulen H, Zhao M, Jin L, Stomph TJ. 2008. Does increased zinc uptake enhance grain zinc mass concentration in rice? *Ann Appl Bio.***153:**135–147.

Johnson AA, Kyriacou B, Callahan DL, Carruthers L, Stangoulis J, Lombi E, Tester M. 2011. Constitutive overexpression of the OsNAS gene family reveals single-gene strategies for effective iron- and zinc-biofortification of rice endosperm. *PLoS ONE.* **6:**e24476.

Juliano BS 1993. Rice In Human Nutrition. IRRI, Los Banos.

Karmakar A. Bhattacharya S, Sengupta S, Ali N, Sarkar SN, Datta K, Datta SK. 2019. RNAi-mediated silencing of ITPK gene reduces phytic acid content, alters transcripts of phytic acid biosynthetic genes, and modulates mineral distribution in rice seeds. *Rice Sci* **27:**315-328

Kassebaum NJ, Jasrasaria R, Naghavi M, Wulf SK, Johns N, Lozan R, Regan M, Weatherall D, Chou DP, Eisele TP, Flaxman SR, Pullan RL, Brooker SJ, Murray CJL. 2014. A systematic analysis of global anemia burden from 1990 to 2010. *Blood* **123:** 615–624.

Kim SA, Lou GM 2007. Mining iron: Iron uptake and transport in plants. *FEBS Letters* **581:** 2273-2280.

Kim SA, Punshon T, Lanzirotti A, Li L, Alonso JM, Ecker JR, Kaplan J, Lou GM. 2006. Localization of iron in Arabidopsis seed requires the vacuolar membrane transporter VIT1. *Science*, **314:**1295-1298.

Kim SI, Andaya CB, Newman JW, Goyal SS, Tai TH. 2008. Isolation and characterization of a low phytic acid rice mutant reveals a mutation in the rice orthologue of maize MIK. *Theor Appl Genet*. **117:**1291–1301.

Kobayashi T, Nagasaka S, Senoura T, Itai RN, Nakanishi H, Nishizawa NK. 2013. Iron-binding haemerythrin RING ubiquitin ligases regulate plant iron responses and accumulation. *Nat. Commun*. **4:**2792

Kobayashi T, Nakanishi H, Takahashi M, Kawasaki S, Nishizawa NK, Mori S 2001. Vivo evidence that Ids3 from Hordeum Vulgare encodes a dioxygenase that converts 2′-deoxymugineic acid to mugineic acid in transgenic rice. *Planta*. **212**(5-6):864-871.

Koike S, Inoue H, Mizuno D, Takahashi M, Nakanishi H, Mori S, Nishizawa NK 2004. OsYSL2 is a rice metal-nicotianamine transporter that is regulated by iron and expressed in the phloem. *The Plant J* **39:**415-424.

Kok AD, Yoon LL, Sekeli R, Yeong WC, Yusof ZNB, Song LK 2018. Iron biofortification of rice: Progress and prospects. In Rice Crop—Current Developments; Saha F, Khan ZH, Iqbal A, Eds: Intech Open: London, UK, Volume 3, pp. 25–44.

Korshunova YO, Eide D, Clark WG, Guerinot ML, Pakrasi HB 1999. The IRT1 protein from Arabidopsis thaliana is a metal transporter with a broad substrate range. *Plant Mol. Biol.* **40:**37–44

Krishnan S, Datta K, Parkhi, Datta SK 2009. Rice caryopsis structure in Relation to distribution of micronutrients (iron, zinc, β-carotene) of rice cultivars including transgenic indica rice. *Plant Sci.* **177:** 557–562.

Krishnaswami K (1998). Nutritional disorders-old and changing. Lancet 351:1268–1269

Kumar S, Rajagopalan S, Sarkar P, Dorward DW, Peterson ME, Liao HS, Guillermier C, Steinhauser ML, Vogel SS, Long EO 2016. Zinc-induced polymerization of killer-cell Ig-like receptor into filaments promotes its inhibitory function at cytotoxic immunological synapses. *Mol Cell* **62:**21–33

Larson SR, Rutger JN, Young KA, Raboy V 2000. Isolation and genetic mapping of a non-lethal rice (*Oryza sativa* L.) low phytic acid 1 mutation. *Crop Sci.,* **40:**1397–1405

Lee S, Jeon US, Lee SJ, Kim YK, Persson DP, Husted S, Schjorring JK, Kakei Y, Masuda H, Nishizawa NK, G (2009) An, Iron fortification of rice seeds through activation of the nicotianamine synthase gene, *Proc Natl Acad Sci.*,106: 22014–22019.

Lee S, Kim Y-S, Jeon US, Kim Y-K, Schjoerring JK, An G 2012. Activation of rice nicotianamine synthase 2 (OsNAS2) enhances iron availability for biofortification. *Mol Cells,* **33**(3):269-275.

Leitenmaier B, Küpper H 2013. Compartmentation and complexation of metals in hyperaccumulator plants. *Front Plant Sci* **4:**374–380

Levenson, CW, Morris D. 2011. Zinc and neurogenesis: Making new neurons from development to adulthood. *Adv Nutr*. 2:96–100.

Liu H, Gan W, Rengel Z, Zhao P 2016. Effects of zinc fertilizer rate and application method on photosynthetic characteristics and grain yield of summer maize. *J Soil Sci Plant Nutr* **16:**550–562

Lucca P, Hurrelland R, Potrykus I 2001. Genetic engineering approaches to improve the bioavailability and the level of iron in rice grains. *Theor Appl Genet.,* **102:**392–397

Lucca P, Poletti S, Sautter C 2006. Genetic engineering approaches to enrich rice with iron and vitamin A. *Plant Physiol.* **126:**291–303

Majumder S, Datta K, Datta SK. 2019. Rice Biofortification: High Iron, Zinc, and Vitamin-A to Fight against "Hidden Hunger. *Agronomy* **9:** 803

Mao H, Wang J, Wang Z, Zan Y, Lyons G, Zou C. 2014. Using agronomic biofortification to boost zinc, selenium, and iodine concentrations of food crops grown on the loess plateau in China. *J Soil Sci Plant Nutr* **14:**459–470.

Maret W. 2013. Zinc biochemistry: from a single zinc enzyme to a key element of life. Adv Nutr Int Rev J 4:82–91.

Masuda H, Kobayashi T, Ishimaru Y, Takahashi M, Aung MS, Nakanishi H, Mori S Nishizawa NK 2013. Iron-biofortification in rice by the introduction of three barley genes participated in mugineic acid biosynthesis with soybean ferritin gene. *Front Plant Sci* **4:**132.

Masuda H, Suzuki M, Morikawa KC, Kobayashi T, Nakanishi H, Takahashi M, Saigusa M, Mori S, Nishizawa NK 2008. Increase in Iron and zinc concentrations in rice grains via the introduction of barley genes involved in phytosiderophore synthesis. *Rice* **1:**100-108

Masuda H, Usuda K, Kobayashi T, Ishimaru Y, Kakei Y, Takahashi M, Higuchi K, Nakanishi H, Mori S, Nishizawa NK 2009. Overexpression of the barley nicotianamine synthase gene HvNAS1 increases iron and zinc concentrations in rice grains. Rice 2:155-166

Mayer JE, Potrykus I 2011. Golden Rice' and biofortification—their potential to save lives is being hampered by overzealous regulation. *ActaHortic,* **941:**21–34.

McGrath JM, Lobell DB 2013. Reduction of transpiration and altered nutrient allocation contribute to nutrient decline of crops grown in elevated CO2 concentrations. *Plant Cell Environ* **36:** 697–705.

Meng FH, Liu D, Yang XE, Shohag MJI, Yang JC, Li TQ, Lu L. Feng Y 2014. Zinc uptake kinetics in the low and high affinity systems of two contrasting rice genotypes. *J Plant Nutr Soil Sci.* **177:**412–420.

Milner MJ, Seamon J, Craft E, Kochian LV 2013. Transport properties of members of the ZIP family in plants and their role in Zn and Mn homeostasis. *J Exp Bot*. 64:369–381.

Morgounov A, Gómez-Becerra HF, Abugalieva A, Dzhunusova M, Yessimbekova M, Muminjanov H, Zelenskiy Y, Ozturk L, Cakmak I 2007. Iron and zinc grain density in common wheat grown in Central Asia. Euphytica 155:193–203

Morrissey J, Baxter IR, Lee J, Li L, Lahner B, Grotz N, Kaplan J, Salt DE, Guerinot ML 2009. The ferroportin metal efflux proteins function in iron and cobalt homeostasis in Arabidopsis. *Plant Cell.* **21**(10):3326-38.

Morrissey J, Lou GM. 2009 Iron uptake and transport in plants: The good, the bad, and the ionome. *Chemical Reviews*. **109**:4553-4567

Myers SS, Zanobetti A, Kloog I, Huybers P, Leakey ADB, Bloom A, Carlisle E, Dietterich LH, Fitzgerald G, Hasegawa T, HolbrookNM, Nelson RL, Ottman MJ, Raboy V Sakai H Sartor KA, Schwartz J, Seneweera S, Tausz M,Usui Y. 2014. Rising CO_2 threatens human nutrition. *Nature* **510**:139–142

Nakandalage N, Nicolas M, Norton RM, Hirotsu N, Milham P, Seneweera S 2016. Improving rice zinc biofortification success rates through genetic and crop management approaches in a changing environment. *Front Plant Sci*. 7:764.

Nattinee P, Cakmak I, Panomwan B, Jumniun W, Benjavan R 2009. Role of Zn fertilizers in increasing grain Zn concentration and improving grain yield of rice. In: The Proceedings of the International Plant Nutrition Colloquium XVI. Department of Plant Sciences, UC Davis USA.

Nishiyama R, Kato M, Nagata S, Yanagisawa S, Yoneyama T 2012. Identification of Zn-nicotianamine and Fe-2'-Deoxy mugineic acid in the phloem sap from rice plants (*Oryzasativa* L.). Plant Cell Physiol53:381–390

Ogo Y, Itai RN, Kobayashi T, Aung MS, Nakanishi H, Nishizawa NK 2011. OsIRO2 is responsible for iron utilization in rice and improves growt and yield in calcareous soil. *Plant Mol Biol*. **75**:593–605.

Olsen LI, Palmgren MG 2014. Manyriverstocross: the journey of zinc from soil to seed. *Front Plant Sci* **5:**30

Ozturk L Yazici MA, Yuce C, Torun A, Cekic C, Bagci A, Ozkan H, Braun HJ, Sayers Z, Cakmak I 2006. Concentration and localization of zinc during seed development and germination in wheat. *Physiol Plant* **128**:144–152

Paine JA, Shipton CA, Chaggar S, Howells RM, Kennedy MJ, Vernon G *et al*. 2005. Improving the nutritional value of Golden Rice through increased pro-vitamin A content. *Nat. Biotech.* **23**: 482–487.

Palmgren MG, Clemens S, Williams LE, Kramer U, Borg S, Schjorring JK, Sanders D 2008. Zinc biofortification of cereals: problems and solutions, *Trends Plant Sci* **13**: 464–473

Papoyan A, Kochian LV 2004. Identification of Thlaspica erulescens genes that may be involved in heavy metal hyper accumulation and tolerance. Characterization of a novel heavy metal transporting ATPase. *Plant Physiol*. **136:**3814–3823

Paul S, Ali N, Datt SK, Datta K. 2014. Development of an iron enriched high-yieldings indica rice cultivar by introgression of a high-iron trait from transgenic iron-biofortified rice. *Plant Foods Hum Nutr*. **69**:203–208.

Paul S, Ali N, Gayen D, Datta SK, Datta K 2012. Molecular breeding of Osfer2 gene to increase iron nutrition in rice grain. *GM Crops Food* 3:310–316

Phattarakul N, Rerkasem B, Li LJ, Wu LH, Zou CQ, Ram H, Sohu VS, Kang BS, Surek H, Kalayci M, Yazici A, Zhang FS, Cakmak I 2012. Biofortification of rice grain with zinc through zinc fertilization in different countries. *Plant and Soil* **361**:131-141

Potrykus I 2001. The golden rice "Tale". *In Vitro Cell Dev Biol Plant* **37**:93–100.

Prasad R, Shivay YS, Kumar D. 2014. Agronomic biofortification of cereal grains with iron and zinc. *Adv Agro* **125**: 55–91.

Qu LQ, Yoshihara T, Ooyama A, Goto F, Takaiwa F 2005. Iron accumulation does not parallel the high expression level of ferritin in transgenic rice seeds. *Planta* **222**:225–233.

Ramesh SA, Choimes S, Schachtman D 2004. Over-expression of an Arabidopsis zinc transporter in Hordeum vulgare increases short-term zinc uptake after zinc deprivation and seed zinc content. Plant Mol Biol 54:373–385.

Ramesh SA, Shin R, Eide DJ, Schachtman P 2003. Di_erential metal selectivity and gene expression of two zinc transporters from rice. *Plant Physiol* **133**: 126–134.

Sadeghzadeh B 2013. A review of Zn nutrition and plant breeding. *J Soil Sci Plant Nutr*. 13:905–927.

Sadeghzadeh B, Rengel Z 2011. Zn in soils and crop nutrition. In: Hawkesford MJ, Barraclough P, editors. The molecular basis of nutrient use efficiency in crops. London: Wiley; p. 335–376.

Saltzman A, Birol E, Oparinde A, Andersson MS, Asare-Marfo D, Diressie MT, Gonzalez C, Lividin K, MoursiM, Zeller M. 2017. Availability, production, and consumption of crops biofortified by plant breeding: Current evidence and future potential. *Ann N Y Acad Sci* **139**:104-114.

Sandmann G. 2001. Carotenoid biosynthesis and biotechnological application. *Arch. Biochem. Biophys*. **385**: 4– 12.

Sandmann, G. 1994. Phytoene desaturase genes, enzymes and phylogenetic aspects. *J. Plant Physiol.* **143**: 444– 447.

Sasaki A, Yamaji N, Mitani-Ueno N, Kashino M, Ma JF 2015. Anode localized transporter OsZIP3 is responsible for the preferential distribution of Zn to developing tissues in rice. Plant J84:374–384.

Satoh-Nagasawa N, Mori M, Nakazawa N, Kawamoto T, Nagato Y, Sakurai K, Takahashi H, Watanabe A, Akagi H 2012. Mutations in rice (*Oryzasativa*) heavy metal ATPase 2 (OsHMA2) restrict the translocation of zinc and cadmium. *Plant Cell Physiol*, 53:213–224.

Schaaf G, Ludewig U, Erenoglu BE, Mori S, Kitahara T, von Wirén N 2004. ZmYS1 functions as a proton-coupled symporter for phytosiderophore- and nicotianamine-chelated metals. J Biol Che 279:9091-9096.

Schroeder JI, Delhaize E, Frommer WB, Guerinot M, Lou HMJ, Herrera-Estrella L, Horie T, Kochian LV, Munns R, Nishizawa NK, Tsay, Sanders D 2013. Using membrane transporters to improve crops for sustainable food production. *Nature,* **497**:60-66

Seneweera S 2011. Reduced nitrogen nallocation to expanding leaf blades suppresses ribulose-1,5-bisphosphate carboxylase/oxygenase synthesis and leads to photosynthetic acclimation to elevated CO2 in rice. *Photosynthetica* **49**:145–148

Shivay YS, Prasad R, Pal M 2015. Effects of source and method of zinc application on yield, zinc biofortification of grain and Zn uptake and use efficiency in chickpea (*Cicerarietinum* L.). Commun Soil Sci Plant Anal46:2191–2200.

Sperotto RA, Ricachenevsky FK, Waldow V de A, Fett JP 2012. Iron biofortification in rice: It's a long way to the top. Plant Sci 190:24-39.

Stomph TJ, ,Jian W, Struik PC 2009. Zinc biofortification of cereals :rice differs from wheat and barley. Trends Plant Sci14:123–124.

Stomph TJ, Hoebe N, Spaans E, Van der Putten PEL 2011. The relative contribution of post-flowering uptake of Zn to rice grain Zn density. 3rd International Zn Symposium; 2011 Oct 10–14; Hyderabad, India.

Suzuki M, Takahashi M, Tsukamoto T, Watanabe S, Matsuhashi S, Yazaki,J, Kishimoto N, Kikuchi S, Nakanishi H, Mori S, Nishizawa NK 2006. Biosynthesis and secretion of mugineic acid family phyto siderophores in zinc-deficient barley. *Plant J* **48**:85–97.

Takahashi R, Bashir K, Ishimaru Y, Nishizawa NK, Nakanishi H 2012. The role of heavy metal ATPases, HMAs,in Zinc and Cadmium transport in rice. *Plant Signal Behav*. **7**:1605–1607

Talke IN, Hanikenne M, Krämer U 2006. Zn-dependent global transcriptional control, transcriptional de-regulation and higher gene copy number for genes in metal homeostasis of the hyper accumulator Arabidopsis halleri. *Plant Physiol*. **142**:148–167.

Tan S, Han R, Li P, Yang G, Li S, Zhang P, Wang WB, Zhao WZ, Yin LP 2015. Over-expression of the MxIRT1 gene increases iron and zinc content in rice seeds. *Transgenic Res* 24:109–122.

Tang G, Qin J, Dolnikowski GG, Russell RM, Grusak MA. 2009. Golden Rice is an effective source of vitamin A. *Am. J. Clin. Nutr.* **89**: 1776–1783.

Theil EC. 2003. Ferritin: At the crossroads of iron and oxygen metabolism. *J Nutr* **133**:1549S-1553S.

Theil EC 2011. Iron homeostasis and nutritional iron deficiency. *J Nutr* **141**: 724S-728S

Thomine S, Lelièvre F, Debarbieux E, Schroeder JI, Barbier-Brygoo H 2003. AtNRAMP3, a multispecific vacuolar metal transporter involved in plant responses to iron deficiency. *Plant J.***34**:685–695

Uauy C, Brevis JC, Dubcovsky J 2006. The high grain protein content gene Gpc-B1 accelerates senescence and has pleiotropic effects on protein content in wheat. *J Exp Bot.* **57**:2785–2794.

Uauy C, Distelfeld A, Fahima T, Blechl A, Dubcovsky J 2006. A NAC gene regulating senescence improves grain protein, zinc, and iron content in wheat. *Science* **314:**1298–1301

Vasconcelos M, Datta K, Oliva N, Khalekuzzaman M, Torrizo L, Krishnan S, Oliveira M, Goto F, Datta K 2003. Enhanced iron and zinc accumulation in transgenic rice with the ferritin gene. *Plant Sci* **164**:371–378.

Velu G, Ortiz-Monasterio I, Cakmak I, Hao Y, Singh RP 2014. Biofortification strategies to increase grain zinc and iron concentrations in wheat. *J Cereal Sci* **59**:365–372.

Von Wirén N, Khodr H, Hider RC 2000. Hydroxylated phytosiderophore species possess an enhanced chelate stability and affinity for iron(III). *Plant Physiol* **124**(3):1149-1158.

Von Wirén N, Klair S, Bansal S, Briat J-F, Khodr H, Shioiri T, Leigh RA, Hider RC 1999. Nicotianamine chelates both FeIII and FeII. Implications for metal transport in plants. *Plant Physiol.* **119**:1107–1114

Welch RM 2002. Breeding strategies for biofortified staple plant foods to reduce micronutrient malnutrition globally. *J Nutr.* **132**:495S-499S

White PJ, Broadley MR 2009. Biofortification of crops with seven mineral elements often lacking in human diets - iron, zinc, copper, calcium, magnesium, selenium and iodine. *New Phytolo* **1**:49-84

Wirth J, Poletti S, Aeschlimann B, Yakandawala N, Drosse B, Osorio S, Tohge T, Ferni AR,Günther D, Gruissem W,Sautte C 2009. Rice endosperm iron biofortification by targeted and synergistic action of nicotianamine synthase and ferritin. *Plant Biotechnol J* **7**: 631-644

Yang Q, Zhang C, Chan M, Zhao D, Chen J, Wang Q, Li Q, Yu H, Gu M,Sun S, Liu Q. 2016. Biofortification of rice with the essential amino acid lysine: Molecular characterization, nutritional evaluation, and field performance. *J Exp Bot* **14:**4285-4296

Yang X, Huang J, Jiang Y, Zhang HS. 2007. Cloning and functional identification of two members of the ZIP (Zrt, Irt-like protein) gene family in rice (*Oryza sativa* L.). *Mol Biol Rep* **36:**281–287

Ye X, Al-Babili S, Klöti A *et al.* 2000. Engineering the provitamin A (beta-carotene) biosynthetic pathway into (carotenoid-free) rice endosperm. Science. 287(5451): 303–305.

Yoneyama T, Ishikawa S, Fujimaki S. 2015. Route and regulation of zinc, cadmium, and iron transport in rice plants (*Oryzasativa* L.) during vegetative growth and grain filling metal transporters metal speciation grain Cd reduction and Zn and Fe biofortification. Int J Mol Sci 16:19111–19129.

Yuan L, Wu L, Yang C, Lv Q. 2013. Effects of iron and zinc foliar applications on rice plants and their grain accumulation and grain nutritional quality. J Sci Food Agr 93:254-261

Zeigler RS 2014. Biofortification: vitamin a deficiency and the case for golden rice. Plant Biotechnol: 245–262

Zheng L, Cheng Z, Ai C, Jiang X, Bei X, Zheng Y, Glehn RP,. Welch RM, Miller DD, Lei XG, Shou H. 2010. Nicotianamine, a novel enhancer of rice iron bioavailability to humans. PLoS One 5: e10190

5

Diversity of Micronutrients in Grains, Its Genetic Basis, and Improvement Through Breeding Techniques

Introduction

Rice contains essential micronutrients such as iron (Fe), zinc (Zn), copper (Cu), calcium (Ca), manganese (Mn), magnesium (Mg), etc. which are generally lower when compared with other staple crops like wheat, maize, legumes, and tubers (Adeyeye *et al.*, 2000). The most crucial micronutrients are iron and zinc, whose inadequacy is a leading cause of malnutrition. In this chapter we discuss estimation methodologies and diversity of Fe and Zn, its genetic basis, and improvement for micronutrient content through breeding techniques.

Procedure of Estimation of Fe and Zn Content in Rice Grain

Iron and zinc concentrations in rice samples were generally estimated by colorimetric method Atomic Absorption Spectrometry (AAS), X-ray Fluorescence (XRF), and Inductively Coupled Plasma Optical Emission Spectrometry (ICP-OES). X-Ray Fluorescence (XRF) Spectrometry is extremely helpful in the non-destructive estimation of relative Zn and Fe concentration in rice breeding lines and large germplasm collection (Paltridge *et al.*, 2012). The majority of biofortification programs employ XRF as a metal analysis tool. For estimation of Fe and Zn, panicles of middle row plants of each replication were harvested, sun-dried for 4-5 days to reduce the moisture content to 11-12%, and then seeds were dehusked. Each brown rice sample of about 5g was subjected to energy dispersive X-ray fluorescent spectrophotometer (ED-XRF) as per standard protocol (Stangoulis and Sison 2008, Rao *et al.*, 2014) and expressed in ppm ($\mu g\ g^{-1}$).

Diversity of Fe and Zn Content in Cultivated and Wild Rice

Selecting genotypes with high efficiency of Fe and Zn accumulation from existing germplasm collection could be a realistic approach to deliver the direct benefit of micronutrient nutrition to farmers and consumers and tools for the breeders for developing high-yielding nutrient-rich rice (Dikshit *et al*., 2016). Indigenous germplasm or landraces are considered the transitional group between wild and cultivated varieties (Pusadee *et al*., 2009) and are reported to be rich sources of iron (Fe) and zinc (Zn) content. Brar (2011) surveyed 220 rice genotypes for Fe and Zn content and reported that *indica* and aromatic rice varieties have high Fe and Zn content. Enhancement of Fe and Zn content in rice grain was found possible by utilizing elite landraces as donors in the rice breeding programme (Swamy *et al*., 2016). It was reported that the practice of consuming polished rice grains in Asian populations aggravates malnutrition (Shi *et al*., 2019). Brown rice is proved to be nutritionally more advantageous than polished rice. Saleh *et al*. (2019) found that in un-polished rice Fe and Zn contents varied from 8.45-45.85 and 17.30-41.45 ppm, respectively. Gregorio (2000) has screened rice lines in IRRI germplasm for high iron and zinc content. Among collected 159 rice germplasm from different agro-climatic regions of India considerable variation was observed in the micronutrient density by Maganti *et al*. (2020). They found that iron and zinc concentrations varied from 6.9 to 22.3 mg/kg and 14.5 to 35.3 mg/kg in unpolished, brown rice. Among a subset of 1,138 samples analyzed, iron concentrations ranged from 6.3 to 24.4 µg/g, and for zinc, the range was 15.3 to 58.4 µg/g. A higher range of variation (Fe:8.3-52.15; Zn:3.0-52.7 ppm) was reported by Tripathy (2020) in germplasm and breeding lines, and Vanlalsanga *et al*. (2019) in germplasm from North East India (Fe: 11.42 - 215.62 ppm; Zn:17.98 - 75.8 ppm). Roy and Sharma *et al*. (2014) also found higher diversity in zinc content (0.85- 195.3 ppm) in rice landraces of West Bengal and adjoining areas. Several studies suggested that Zn content in the grain seemed to be more constant than Fe content. Such higher variability in Fe content compared to Zn content was also observed by Maganti *et al*. (2020). It was reported that wild rice, landraces, and *aus* accessions are excellent sources of micronutrients, having a richer concentration of nutrients than cultivated rice (Cheng *et al*., 2005; Banerjee *et al*., 2010; Descalsota-Empleo *et al*., 2019a, b). Genetically, *aus* accession rice is more similar to widely cultivated *indica* rice varieties, making it easier for the breeding programs to utilize this in improving Zn content in modern rice genotypes. Kaliboro, Jamir, UCP122, DZ193, and Khao ToT Long 227 are some examples of *aus* accessions that have higher concentrations of Zn in grain (Norton *et al*., 2014; Descalsota *et al*., 2018). A*us* germplasm was efficiently used in breeding programs at IRRI. The accessions of 3K Genome Project, MAGIC-derived lines (Multi-parent Advanced Generation Inter-Cross), and wild rice introgression lines were analyzed to find potential donors of grain Zn utilized for genetic dissection studies (Bandillo *et al*., 2013; Swamy *et al*., 2018a; Descalsota *et al*., 2018; Zaw *et al*., 2019).

A few germplasm lines consisting of landraces, breeding lines, and varieties selected from 16 national institutes were assessed for Zn content at IIRR using the XRF facility and it was found to be in a range of 7.3 to 52.7 mg/kg. The average Zn content in the brown rice of germplasm varied from 15.9 to 27.3 mg/kg. The Zn content in the polished rice grain of eight extensively cultivated varieties ranged from 16.0 to 26.5 mg/kg (23), and in 246 germplasm lines, it was around 33 mg/kg in polished rice (26). Across the examined sample, about 19% of the mean percentage loss of Zn content was noted during polishing which is 1.9 mg/kg in each 10mg/kg of brown rice. In various germplasm and mapping populations, the mean percentage loss of Zn content was from 11.1 to 27.9% (Figure 1). Reports of zinc content loss as a result of polishing varied from 5 to 30% (26,27). As per the previous reports, this variation might be due to the varying thickness of the aleurone layer amongst rice genotypes (23, 28). The zinc concentration varied from < 5 to 25 mg/kg in the brown rice of commonly cultivated varieties by farmers (16, 20). Whereas, in the polished rice grains of popular rice cultivars extensively grown by farmers, the zinc content was in a range of <12 to 14 mg/kg. Given the Harvestplus threshold value of 28 mg/kg and 19% mean zinc loss, while polishing, germplasm in brown rice with a zinc level of 35 mg/kg is thought to be an excellent source for meeting the targeted zinc content.

Trait Association with Micronutrient Contents

Sanghamitra *et al*. (2022) found that Zn content was negatively correlated with panicle exertion, however, the grain yield and panicle exertion were positively correlated. Several studies indicated that expression of metal homeostatic genes such as *OsNAC, OsYSL2, OsYSL9, OsZIP4*, etc. in the flag leaf was highly correlated with the high grain Zn content (Banerjee and Chandel, 2011; Swamy *et al*., 2016). But under low panicle exertion, the concentration of Zn might be higher in flag leaf which possibly increased the Zn content in fertile spikelet through transmission. Sanghamitra *et al* (2022) found that Fe content had a significant positive correlation with the Zn content of the grain and both Fe and Zn content was found to have a positive association with grain size. Such a positive association of Fe with Zn was identified by several researchers (Anuradha *et al*., 2012; Bollinedi *et al*., 2020; Maganti *et al*., 2020). A positive association of Zn and Fe content was evident by the co-location of QTLs for

Fe and Zn in brown rice (Swamy *et al.*, 2016), In addition to this, Zn content was found to have a significant positive association with panicle length and negative association with days to 50% flowering. Grain Fe and Zn contents not only depend on the genotypes but also depend on the environment and interaction of genotype and environment (Gregorio, 2002; Chandel *et al.*, 2010). Loamy soil, neutral PH, and higher organic matter content were reported to be responsible for higher Fe and Zn content (Chandel *et al.*, 2010). One of the most significant attributes of high Zn rice development is the correlation between Zn content in grain and grain yield. Numerous studies have shown a strong inverse relationship between rice yield and grain zinc content (Norton *et al.*, 2010). While a positive association was noted only in Zn-deficient soil (Gregorio, 2002). Grain zinc content is a multifaceted trait, that has been shown to be highly impacted by climate and external soil conditions (Chandel *et al.*, 2010; Anuradha *et al.*, 2012; Swamy *et al.*, 2016; Naik *et al.*, 2020).

Sanghamitra *et al.* (2022) found that since red grain genotypes were predominant in the Nayagarh district in Odisha, so red grain genotypes might be responsible for higher Fe content in that district. Whereas the red-laterite category of soil and rich mineral base of Deogarh and Mayurbhanj districts (Sahu and Mishra, 2005; Mishra and Nanda, 2008) might be responsible for generating landraces with greater ability for uptake and accumulation in grain resulting in higher Zn content of this region. Meteorological parameters such as temperature, relative humidity, and rainfall; soil factors like nutrient status, organic matter, and pH; and agronomic practices such as irrigation, tilling, fertilizer application, and cultivation system (White and Broadley 2009; Joshi *et al.*, 2010; Chandel *et al.*, 2010) should be taken into consideration. According to an investigation carried out by Wissuwa *et al.* (2008), the native zinc in soils, genotypes, and the applied zinc fertilizer all had a significant impact on grain zinc concentration. According to Wang *et al.* (2014), ZnSO4 application along with proper water management in soil by alternate wetting and drying method (AWD) improved grain yield in addition to a higher Zn content in rice grain. Pandian *et al.* (2011) performed a field trial at three locations with 17 rice genotypes. The result revealed that zinc concentration in grain differed dramatically between genotypes and locations. Thus, G×E analysis is required to assess potential germplasm and the stability of mineral accumulation over multiple generations and test sites (Gregorio 2002; Wissuwa *et al.*, 2008; Impa *et al.*, 2013; Naik *et al.*, 2020). Hence, the Zn-biofortified genotypes must be stable in terms of grain Zn content as well as grain yield before being commercially released as varieties.

Identification of QTLs for Micronutrient Content in Rice

In the last twenty years, over 250 QTLs have been found to be associated with zinc and iron content and mapped on 12 chromosomes in the rice genome using mapping populations originating from various intraspecific and interspecific crosses. The overview of QTLs linked with zinc and iron contents and the flanking markers are presented in **Table 1**.

Table 1. List of important QTLs and associated probable functional genes for grain Fe and Zn content in rice

Population type	Cross	No.of total QTLs	Chromosomes/ Chromosome arms	PVE range (additive effect QTLs)	Important QTLs/associated functional genes	References
DH	IR6 (*indica*) x Azucena (*japonica*)	Zn-2, Fe-3	1,12, 2,8,12	-		Stangoulis et al. 2007
RILs	Zhengshan97 (*indica*) x Minghui63 (*indica*)	Zn-3, Fe-2	5,7,11,1,9	5.3-18.61, 11.11-25.81		Lu et al. 2008
ILs	*O.sativa* ssp. *Indica* (Teqing) x *O. rufipogon* Griff.	Zn-2; Fe-1	5,8;2	5-11;7	qZn8.1	Garcia-Oliveira et al. 2009
RILs	Bala (*indica*) x Azucena (*japonica*)	Zn-4; Fe-4	6,7,10; 1,3,4,7	11.2-14.8; 9.7-21.4		Norton et al. 2010
DH	ZYQ8 (*indica*) x JX17 (*indica*)	Zn-2	4,6	10.83-12.38		Zhang et al. 2011
RILs	Madhukara (*indica*) x Swarna (*indica*)	Zn-6; Fe-7	3,7,12;1, 5,7,12	29.35; 69-71	OsYSL1 and OsMTP1 for Fe and OsARD2, OsIRT1, OsNAS1, and OsNAS2 for Zn	Anuradha et al. 2012
F2 population	PAU201 (*indica*) x Palman 579 (*indica*)	Zn-3; Fe-8	2,10;2,3 ,7,10,12	4.7-19.1; 2.4-26.8		Kumar et al. 2014
RILs	Swarna (*indica*) x Moroberekan (*japonica*)	Fe-1	1	39	qFe1.1	Indurkar et al. 2015
BILs	Ce258 (*indica*) x IR75862 (*japonica*) and ZGX1 (*indica*) x IR75862 (*japonica*)	Zn-4; Fe-1	3,6,7,8; 6,11	2-24.4; 10.2-18.3		Xu et al. 2015
BILs	*O.sativa* (XB)x *O.rufipogon* (accession of DWR)	Zn-6; Fe-3	3,4,6,7,10,12; 3,6,9	5.3-11.8; 6.1-28.2		Hu et al. 2016
BILs	Nipponbare (*japon-*	Zn-4	2,9,10	15.0-21.9	qGZn9a, qGZn9b	Ishikawa et al. 2017

BIL	Swarna/ IRGC81832 (*O. Nivera*) and Swarna/ IRGC81848 (*O. nivera*)	Zn-13, Fe-17	1, 2, 3, 5, 6, 8, 9, 10, 12	3-36	qFe2.1, qFe3.1, qFe8.2 and qZn12.1 were consistent Associated genes: OsNAS2, OsZIP4, Myb transcription factor	Swamy et al. 2018 a
DH	PSBRc82 (*indica*) X Joryeongbyeo (*japonica*) and PSBRc82 (*indica*) x IR69428 (*indica*)	Zn-8; Fe-1	2,3,6,8,11,12; 4	7.5-22.8; 9.4	OsNRAMP, OsNAS, OsZIP, OsZIFL	Swamy et al. 2018b
MAGIC	Association panel	Zn- 7	1, 2, 4, 6, 7, 12	9.2- 13.75	Zn7:1, OsMTP6, OsNAS3, OsMT2D, OsVIT1, and OsNRAMP7	Descalsota et al. 2018
Genome-wide association analysis	Association panel (152 coloured rice accessions)	Zn-5	1, 6, 12	11.9-17.9	qZn12.2 . OsHMA9, OsMAPK6, OsNRAMP7, OsMADS13, and OsZFP252	Descalsota- Empleo et al. 2019a
DH	IR 64 (*indica*)x IR69428 (*indica*) and BR 29 (*indica*)x IR75862 (*japonica*)	Zn-8	2,3,5,7,8,9,11	8.6-27.7	qZn2.1, qZn5.1, qZn9.1, qZn11.1	Descalsota-Empleo et al. 2019b
F4 population	PAU201 (*indica*) x Palman 579 (*indica*)	Zn-1; Fe-5	6;5,7,9	25; 34.6-95.2		Kumar et al. 2019
MAGIC	Association panel	Zn-3	1, 5, 7	17.57-20.0	OsNAS3	Zaw et al. 2019

Backcross population	RP-Bio226 (*indica*)/ Sampada (*indica*)	four consistent QTLs for Fe () and two QTL for Zn content ()	1, 6	34.2; 17.1%	qFe1.1, qFe1.2, qFe6.1 and qFe6.2 qZn1.1 and qZn6.2 Potential candidate genes: OsPOT, OsZIP4, OsFDR3, OsIAA5	Dixit et al. 2019
DH	IR05F102 (*indica*)/ IR69428 (*indica*)	Zn-4	1, 5, 9, 12	8.96- 15.25	*OsZIP6* on QTL *qZn5.1*	Calayugan et al. 2020
DH	Hwaseonchal (*japonica*) /Goami 2 (*japonica*)	Zn- 14 Fe- 7	1, 2, 4, 5, 6, 7, 9, 10, 11, 12	12.60- 46.80	qFe7 and qZn7	Jeong et al. 2020
RIL	PR116/ Ranbir Basmati	Zn- 6	1, 11, 2, 9	5 - 37.84	major QTL qZPR.1.1 (PV 37.84%)	Suman et al. (2021).
BIL	Hwaseong (*japonica*)/ IRGC105491 (*Oryza rufipogon*)	Zn-3 Fe- 3	6, 8, and 10		qFe10 and qZn10; transcription factor gene, namely the rice homeobox gene 9 (LOC_Os10g33960)	Adeva et al. 2023

In an initial study, a few numbers of QTLs for mineral (P, Fe, Zn, Cu, and Mn) concentrations in rice grain were identified by Stangoulis *et al.* (2007) among them 2 and 3 QTLs were for Zn and Fe concentration, respectively. 31 potential QTLs for P, K, Ca, Mg, Zn, Fe, Mn, and Cu contents were identified by Garcia-Oliveira *et al.* (2009) using introgression lines originating from the inter-specific cross.

Favourable alleles for most of the QTLs (26 QTLs) including 2 QTLs for Zn content and 1 QTL for Fe content were contributed by wild rice *O. rufipogon*. Norton *et al.* (2010) also mapped 41 QTLs for the concentration of 17 elements in rice grain. In subsequent studies, several QTLs were reported which explained a large amount of PVE for zinc and iron contents (Zhang *et al.*, 2011; Anuradha *et al.*, 2012; Kumar *et al.*, 2014; Indurkar *et al.*, 2015). Anuradha *et al.*, 2012 detected a few major QTLs and associated probable functional genes such as OsYSL1 and OsMTP1 for Fe and OsARD2, OsIRT1, OsNAS1, and OsNAS2 for Zn content. Backcross-derived populations from two *indica/ japonica* cross (Xu *et al.*, 2015) and inter-specific cross (Hu *et al.* 2016) were used to detect QTLs for Fe and Zn content. One accession of *O. meridionalis* was used to generate a backcross population and detect and fine-map a robust genomic locus, qGZn9 (Ishikawa *et al.*, 2017). They suggested using qGZn9b as a favourable allele to breed rice genotypes with a high level of zinc in the grain. With an aim to characterize the impact of QTL interactions on the phenotypic expression of the trait, the interaction among the discovered QTLs for grain zinc was also investigated (Swamy *et al.*, 2018a; Descalsota-Empleo *et al.*, 2019a; Calayugan *et al.*, 2020).

Swamy *et al.* (2018a) employed inter-specific cross to identify consistent QTLs such as qFe2.1, qFe3.1, qFe8.2, and qZn12.1 and associated genes, OsNAS2, OsZIP4, Myb transcription factor. In another study two double haploid (DH) populations were utilized for the detection of genomic region for Fe and Zn content and the probable functional genes inside QTLs region such as OsNRAMP, OsNAS, OsZIP, OsZIFL which might be useful for marker-assisted breeding for this important trait (Swamy *et al.*, 2018b). Additionally, they discovered eight epistatic interactions in a double haploid population for Zn, Cu, Mg, and Na. Descalsota-Empleo *et al.* (2019b) performed a phenotypic study on two DH populations (*indica* x *indica*) across two growing seasons and also genotyped it using a 6 K SNP chip. A total of 50 QTLs were identified for grain nutrient content. Among them, 8 QTLs were found to be associated with grain zinc content explaining a phenotypic variance of 8.6-27.7%. The single-QTL lines with qZn9.1 were found to have the highest grain zinc content with a mean of 18.1 to 19.1 mg/kg in two consecutive seasons respectively. They observed an increase in the Zn concentration in the grain due to an increase in the numbers of QTL and the highest Zn content in grain was found to be 28.2 and 24.3 mg/kg in two growing seasons, respectively in lines with four QTL (qZn2.1 + qZn5.1 + qZn5.1 + qZn11.1). This investigation pointed to the fact the possibility of QTL pyramiding for improving the zinc content in rice. In another study, Kumar *et al.* (2019) detected one QTL for Zn content and five QTLs for Fe content using an F4 population derived from an indica x indica cross (PAU 201/ Palman). Double haploid population derived from indica/ indica cross (Calayugan *et al.* 2020) and japonica/ japonica cross (Jeong *et al.* 2020) were employed to detect robust QTLs such as qZn5.1 and qFe7 and qZn7, respectively. Dixit *et al.* (2019) using RP-Bio226/ Sampada cross detect a few robust QTLs and potential candidate genes such as OsPOT, OsZIP4, OsFDR3, and OsIAA5. Employing inter-specific cross Adeva *et al.* (2023) recently identified consistent QTLs qFe10 and qZn10 and transcriptional factor gene (LOC_Os10g33960). Suman *et al.* (2021) evaluated 190 RILs developed from PR116 and Ranbir Basmati cross in two environments. Two major QTLs, qZPR.1.1 and qZPR.11.1 were identified for grain Zn content in polished rice showing a phenotypic variance of 37.84% and 15.47% respectively. Recently, Descalsota *et al.* (2018) identified seven QTLs on chromosomes 1, 2, 4, 6, 7, and 12 and Zaw *et al.* (2019) discovered three QTLs on chromosomes 1, 5, and 7 using association mapping on diverse panels showing high variation in grain Zn content. Recently, Descalsota *et al.* (2018) identified seven QTLs on chromosomes 1, 2, 4, 6, 7, and 12 and Zaw *et al.* (2019) discovered three QTLs on chromosomes 1, 5, and 7 using association mapping on diverse panels showing high variation in grain Zn content. Recently, Descalsota *et al.* (2018) identified seven QTLs on chromosomes 1, 2, 4, 6, 7, and 12 and Zaw *et al.* (2019) discovered three QTLs on chromosomes 1, 5, and 7 using association mapping on diverse panels showing high variation in grain Zn content. Recently, Descalsota *et al.* (2018) identified seven QTLs on chromosomes 1, 2, 4, 6, 7, and 12 and Zaw *et al.* (2019) discovered three QTLs on chromosomes 1, 5, and 7 using association mapping on diverse panels showing high variation in grain Zn content. Recently, Descalsota *et al.* (2018) identified seven QTLs on chromosomes 1, 2, 4, 6, 7, and 12 and Zaw *et al.* (2019) discovered three QTLs on chromosomes 1, 5, and 7 using association mapping on diverse panels showing high variation in grain Zn content. Recently, Descalsota *et al.* (2018)

identified seven QTLs on chromosomes 1, 2, 4, 6, 7, and 12 and Zaw *et al.* (2019) discovered three QTLs on chromosomes 1, 5, and 7 using association mapping on diverse panels showing high variation in grain Zn content. Recently, Descalsota *et al.* (2018) identified seven QTLs on chromosomes 1, 2, 4, 6, 7, and 12 and Zaw *et al.* (2019) discovered three QTLs on chromosomes 1, 5, and 7 using association mapping on diverse panels showing high variation in grain Zn content. Recently, Descalsota *et al.* (2018) identified seven QTLs on chromosomes 1, 2, 4, 6, 7, and 12 and Zaw *et al.* (2019) discovered three QTLs on chromosomes 1, 5, and 7 using association mapping on diverse panels showing high variation in grain Zn content. Recently, Descalsota *et al.* (2018) identified seven QTLs on chromosomes 1, 2, 4, 6, 7, and 12 and Zaw *et al.* (2019) discovered three QTLs on chromosomes 1, 5, and 7 using association mapping on diverse panels showing high variation in grain Zn content. Recently, Descalsota *et al.* (2018) identified seven QTLs on chromosomes 1, 2, 4, 6, 7, and 12 and Zaw *et al.* (2019) discovered three QTLs on chromosomes 1, 5, and 7 using association mapping on diverse panels showing high variation in grain Zn content. Recently, Descalsota *et al.* (2018) identified seven QTLs on chromosomes 1, 2, 4, 6, 7, and 12 and Zaw *et al.* (2019) discovered three QTLs on chromosomes 1, 5, and 7 using association mapping on diverse panels showing high variation in grain Zn content. Descalsota *et al.* (2018) exmined 144 MAGIC Plus lines for agronomic and biofortification traits for two seasons at two different locations. Additionally, they assessed disease resistance in these lines for one season in screen house. Grain Fe and Zn content was estimated using X-ray fluorescence technology. These lines were genotyped using genotype by sequencing method and a total of 14,242 SNP markers used for association analysis. QTL, qZn7:1 was consistent across all four environemts and qZn6:2 was identified in two enviromments. Metal homeostasis genes such as OsMTP6, OsNAS3, OsMT2D, OsVIT1, and OsNRAMP7 were found to be co-localized with QTLs associated with Zn and Fe. Descalsota-Empleo *et al.* (2019a) using an association panel of 152 coloured rice accessions identified 5loci for Zn content. One robust QTL qZn12.2 and associated probable functional genes OsHMA9, OsMAPK6, OsNRAMP7, OsMADS13, and OsZFP252 were detected in that study.

Meta-QTL analysis delivers consolidated, specific, and narrower confidence ranges for numerous QTLs reported for a trait (Goffinet and Gerber 2000; Arcade *et al.* 2004; Swamy *et al.* 2011). 22 meta-QTLs (rMQTLs) have been identified for grain Zn concentration across ten chromosomes by Jin *et al.* (2015). Raza *et al.* (2019) also performed meta-QTL analysis for grain-zinc content identified from 24 mapping populations and three diverse germplasm panels. They identified 46 MQTLs. Similarly, Raza *et al*. (2019) carried out a meta-QTL analysis of grain-Zn QTLs reported from 24 mapping populations and three diverse germplasm sets and identified 46 MQTLs. Seven meta-QTLs ($rMQTL_{2.1}$, $rMQTL_{4.4}$, $rMQTL_{6.4}$, $rMQTL_{8.2}$, $rMQTL_{8.3}$, $rMQTL_{8.4}$, and $rMQTL_{12.4}$) were found to be common between two studies (Jin *et al*. 2015; Raza *et al*. 2019). In a separate study by Soe (2020), 208 QTLs identified for grain Zn content out of 26 studies were mapped on the consensus map, leading to the identification of 45 MQTLs. These MQTLs exhibited narrower confidence internals compared to the mean values of original QTLs. In addition, the study also identified numerous stable QTLs and the associated makers, which can be beneficial for marker-assisted selection (MAS) programs. Moreover, the precise MQTL regions identified can offer a scope to identify and validate candidate genes.

Breeding Strategies for Fe and Zn Biofortification

Sanghamitra *et al*. (2022) found that among nutrient-rich germplasm, some had moderately high grain yields such as Champeisiali which contained high Fe (44.1 ppm) and Zn (27.39 ppm) over the years along with moderately high grain yield (20 g/plant). Therefore, improvement of Fe and Zn content in rice grain along with high grain yield is possible. Low heritability of these micronutrient traits might be the outcome of the involvement of a large number of minor genes and higher genotype x environment interaction. A skewed distribution was also observed in a RIL population by Banu and Jagadeesh (2014) for grain Zn content, indicating the involvement of several minor genes with duplicate gene interactions indicating inadequate scope of improvement by direct pedigree selection. In the contrary, the bulk-pedigree breeding method, a modified pedigree selection method was proved quite effective for traits with low heritability (Wynne and Gregory 1981). It was observed that initial bulking followed by selection from F_5 onwards through the pedigree method would help in getting a few transgressive segregants, rare combinations, and fixing the desirable genes. It was suggested that high Zn donors with acceptable yield levels could be used in crossing with high-yielding varieties and the selection of desirable high-yielding high Zn segregants from F_4 generation onwards could hasten the process of high-Zn variety development and simultaneous maintenance of yield potential (Swamy *et al*., 2016). Advanced backcross breeding also was recommended to transfer high Zn traits from un-adapted donors to high-yielding popular rice background. Recently released high-yielding varieties with high Zn content such as CR Dhan 311, DRR Dhan 45, DRR Dhan 49, CR Dhan 315, etc. under biofortification breeding programme exercising backcross and bulk-pedigree breeding methods indicated the evidence of successfully combining high yield with high Zn content in rice.

For a successful breeding program, it is essential to identify and systematically characterize genetic variability for any trait and utilize it effectively. In the majority of the elite modern rice cultivars and their nearest elite genetic pools contain lower concentrations of zinc in grain (Gregorio, 2002). On the other hand, *Oryza nivara*, *O. rufipogon*, *O. longistaminata*, and *O. barthii* accessions, landraces, colored rice, and *aus* and aromatic accessions mostly have high concentrations of zinc in grain Swamy *et al.*, 2016, 2018a, b; Ishikawa *et al.* 2017). However, due to poor phenotype and lower grain yield, these accessions may not be ideal. So, systematic pre-breeding efforts are essential to develop rice genotypes with high zinc content as well as favourable agronomic traits.

Utilization of the advanced backcross method is a desirable strategy for effectively using wild rice accessions for genetic dissection and the development of high zinc introgression lines (Balakrishnan *et al.*, 2020). Numerous introgression lines having rich concentrations of zinc in grain derived from wild rice accessions have been reported before (Ishikawa *et al.* 2017; Swamy *et al.*, 2018a). Populations derived from multiple parents used to select for transgressive variants with a combination of desirable traits have successfully produced numerous desirable transgressive variants with higher grain zinc content (Gande *et al.*, 2013; Ishikawa *et al.*, 2017; Descalsota-Empleo *et al.*, 2019a, b). Apart from that several gene-specific markers such as *OsZIP1, OsZIP3, OsZIP4, OsZIP5, OsZIP8, OsNAS1, OsNAS2, OsNRAMP1*, etc. showed a very good association with grain Zn (Chandel *et al.* 2011; Gande *et al.*, 2014). Utilizing marker-assisted breeding methods for zinc-enriched rice, including QTL deployment, QTL pyramiding, and marker-assisted recurrent selection, offers a faster and more precise approach for enhancing rice varieties with high zinc content alongside other traits (Hill *et al.*, 2008; Boyle *et al.*, 2017).

Zn-Biofortified Rice Varieties Released in Different Countries

It was suggested by Swamy *et al.* (2016) that high Zn donors with acceptable yield level could be used in crossing with high yielding varieties and selection of suitable high yielding high Zn segregants from F_4 generation onwards could accelerate the process of developing high yielding biofortified variety with high Zn. Advanced backcross breeding also was suggested to transfer high Zn trait from un-adapted donors to high yielding rice varieties. Recently released high yielding varieties with high Zn content such as CR Dhan 311, DRR Dhan 45, DRR Dhan 48, DRR Dhan 49, DRR Dhan 63, CR Dhan 315, Chhattisgarh Zinc Rice-1, Chhattisgarh Zinc Rice-II, Zinco Rice MS, etc. under biofortification breeding programme exercising backcross, pedigree and bulk-pedigree breeding methods indicated the evidence of successfully combining high yield with high Zn content in rice. Some of these biofortified varieties are as follows.

- ***DRR Dhan 45:*** First high zinc variety notified in India. It is semi-dwarf medium duration (125 days) for irrigated conditions. Zn content is 24.0 ppm in polished rice. It was released for Tamil Nadu, Andhra Pradesh and Karnataka with grain yield of 5.2 t/ha.
- ***Chhattisgarh Zinc Rice-1:*** It has a average 22 ppm Zn content in milled rice. It is semi dwarf (95-100 cm height) in nature and early mature in 105-110 days. It has long bold grains and selected for 'direct seeded rainfed, aerobic and irrigated Ecosystem' of Chhattisgarh plains.
- ***CR Dhan 315 (IET 27179: CR 2826-1-1-2-4B-2-1):*** It is released and notified for the states of Gujarat and Maharastra as biofortified rice variety. It contains average 24.9 ppm zinc in milled rice in zone VI. It has medium duration (125-135 days), semi-dwarf plant type (110 cm) with medium slender grain and good grain quality. It is suitable for irrigated and favorable shallow rainfed ecology. National average of grain yield of this variety is 5054 kg/ha.
- ***Zinco Rice MS:*** It is released and notified for cultivation in irrigated ecology of Chhattisgarh, West Bengal and Odisha. It contains 27 ppm zinc in milled rice. It has medium duration (135 days) with medium slender grain. The average grain yield is 5800 kg/ha.
- ***DRR Dhan 49:*** DRR Dhan 49 is semi-dwarf variety with duration of 130-135 days and has medium slender grain type. It is suitable for the irrigated ecologies. It is a high zinc variety (25.2 ppm) with BLB resistance having genes of xa21+xa13 and moderately tolerant to Blast. It is released in 2018 for the states of Gujarat, Maharashtra and Kerala.
- ***DRR Dhan 48:*** DRR Dhan 48 is a high zinc variety (20.91 ppm) suitable to irrigated ecology. It is semi-dwarf, medium slender variety with duration of 135-140 days. It is a variety resistant to BLB having genes of xa21+xa13+xa5. It has a yield potential of 5-5.5 t/ha and is released for the states of Telangana, Karnataka, Tamil Nadu and Kerala

- ***DRR Dhan 63:*** DRR Dhan 63 is semi-dwarf variety with duration of 130 days and has short bold grain type. It is suitable for the irrigated ecologies. It is a high zinc variety (24.2 ppm) with moderately resistant to leaf and neck blasts, bacterial leaf blight and planthopper. It is released in 2021 for the states of Uttar Pradesh, Odisha and Kerala.
- ***Chhattisgarh Zinc Rice-2:*** It has a average 22 ppm Zn content in milled rice. It is semi dwarf (91 cm height) in nature and early mature in 105-110 days. It has short slender grains and suitable for irrigated and rainfed ecologies in Chhattisgarh plains.

Biofortification programme have resulted in the successful release of several high-Zn rice varieties in several countries of Asia. Five high-Zn rice varieties such as BRRI Dhan 62, BRRI Dhan 64, BRRI Dhan72, BRRI Dhan74 and BRRI Dhan 84 have been released for cultivation in Bangladesh. BRRI Dhan 72 had an average yield of 5.7 tons per hectare in 125 days was found suitable in cropping season with three crops. It contained about 9 percent protein along with 23 milligrams of zinc per kilogram of rice. In Philippines and Indonesia, NSIC Rc 460 and Inapari Nutri Zn, respectively have been released. All these high-Zn rice varieties have higher grain-Zn (>24 ppm) content along with suitable agronomic traits and biotic and abiotic stress tolerance (Swamy *et al.*, 2016). Several promising high-Zn lines have been successfully tested in Myanmar and Cambodia and in some African countries. In addition to that several gene specific markers such as *OsZIP1, OsZIP3, OsZIP4, OsZIP5, OsZIP8, OsNAS1, OsNAS2, OsNRAMP1*, etc. showed a very decent correlation with grain Zn content (Chandel *et al.*, 2011). Marker assisted breeding for high Zn rice using major gene/QTL based markers also could be a faster and meticulous approach.

High iron content rice such as IR 68144, was bred through molecular breeding approaches by overexpressing soybean ferritin gene registered to increase iron content in milled rice grain by 3.7-fold (Vasconcelos *et al.*, 2003). Recently, Paul *et al.* (2014) successfully transferred this trait to Indian mega rice variety Swarna, resulting in 2.54-fold more iron in milled rice grain as compared to control Swarna.

Fig. 1: High zinc biofortified rice varieties (field view, grain and milled rice) in India, a. CR Dhan 315 and b. Zincorice MS

References

Adeva, C.; Yun, Y.T.; Shim, K.C.; Luong, N.H.; Lee, H.S.; Kang, J.W.; Kim, H.-J.; Ahn, S. N. 2023. QTL Mapping of Mineral Element Contents in Rice Using Introgression Lines Derived from an Interspecific Cross. Agronomy, 13, 76. https://doi.org/10.3390/agronomy13010076

Anandan A, Rajiv G, Eswaran R, Prakash M. 2011. Genotypic variation and relationships between quality traits and trace elements in traditional and improved rice (OryzasativaL.) genotypes. J Food Sci. 76:H122– 30. doi: 10.1111/j.1750-3841.2011.02135.x

Anuradha K, Agarwal S, Rao YV, Rao KV, Viraktamath BC, Sarla N. 2012. Mapping QTLs and candidate genes for iron and zinc concentrations in unpolished rice of Madhukar× Swarna RILs. Genetics. **508**: 233

Arcade, A. *et al.* 2004. BioMercator: integrating genetic maps and QTL towards discovery of candidate genes. *Bioinformatics.* **20:** 2324–6.

Babu VR, Neeraja CN, SanjeevaRao D, Sundaram RM, Longvah T, Usharani G, *et al.* 2014. Bioforotification in Rice.DRR Technical Bulletin No. 81/2014. Hyderabad: Directorate of Rice Research. pp. 86.

Balakrishnan, D., Surapaneni, M., Rao, Y. V., Raju, A. K., Mesapogu, S., Beerelli, K., *et al.* 2020. Detecting CSSLs and yield QTLs with additive, epistatic and QTL _ environment interaction efects from Oryza sativa _ O. nivara IRGC81832 cross. *Sci. Rep.* **10:** 7766. doi: 10.1038/s41598-020-64300-0

Banerjee S, Chandel G 2011. Understanding the role of metal homeostasis related candidate genes in Fe/Zn uptake, transport and redistribution in rice using semi-quantitative RT-PCR. J Plant Mol Biol Biotechnol **2**:33–46

Banerjee S, Sharma D, Verulkar S, Chandel G 2010. Use of in silico and semi quantitative RT-PCR approaches to develop nutrient rich rice (*Oryza sativa* L.). *Ind J Biotechnol* 9:203–212

Bollinedi H, Yadav AK, Vinod KK, Gopala KS, Bhowmick PK, Nagarajan M. 2020. Genome-wide association study reveals novel marker trait associations governing the localization of Fe and Zn in the rice grain. *Front genet,* **11**:213

Bandillo N, Raghava C, Muyco PA, Sevilla MAL, Lobina IT, Dilla-Ermita CJ, Tung CW, McCouch S, Thomson M, Mauleon R, Singh RK, Gregorio G, Redoña E, Leung H 2013. Multi-parent advanced generation inter-cross (MAGIC) populations in rice: progress and potential for genetics research and breeding. *Rice* **6:**11

Bouis HE, Saltzman A. Improving nutrition through biofortification: a review of evidence from harvestplus, 2003 through (2016). Glob Food Sec. (2017) 12:49–58. doi: 10.1016/j.gfs.2017.01.009

Brar B, Jain S, Singh R, Jain RK 2011. Genetic diversity for iron and zinc contents in a collection of 220 rice (*Oryza sativa L.*) genotypes. *Indian J Genet.* **71**: 67-73.

Calayugan, M. I. C., Formantes, A. K., Amparado, A., Descalsota-Empleo, G. I, Nha, C. T., Inabangan-Asilo, M. A., *et al.* 2020. Genetic analysis of agronomic traits and grain iron and zinc concentrations in a doubled haploid population of rice (*Oryza sativa* L.). Sci. Rep. 10:2283. doi: 10.1038/s41598-020-59184-z

Chandel G, Banerjee S, See S, Meena R, Sharma DJ, Verulkar SB. 2010. Effects of different nitrogen fertilizer levels and native soil properties on rice grain Fe, Zn and protein contents. *Rice Sci.* 17:213– 27. doi: 10.1016/S1672-6308(09)60020-2

Cheng Z-Q, Huang X-Q, Zhang Y-Z,Qian J, Yang M-Z, Wu C-J, Liu J-F. 2005. Diversity in the Contentof Some Nutritional Components in Husked Seeds of Three Wild Rice Species and Rice Varieties in YunnanProvince of China. *J Integr Plant Biol*, **47**: 1260–1270.

Descalsota-Empleo GI, Amparado A, Inabangan-Asilo MA, Tesoro F, Stangoulis J, Reinke R, *et al*. 2019. Genetic mapping of QTL for agronomic traits and grain mineral elements in rice. *The Crop Journal.*

Descalsota-Empleo GI, Noraziyah AAS, Navea IP, Chung C, Dwiyanti MS, Labios RJD, Ikmal AM, Juanillas VM, Inabangan-Asilo MA, Amparado A,Reinke R. 2019. Genetic dissection of grain nutritional traits and leaf blight resistance in rice. *Genes*, **10**(1): 30.

Dikshit, N, Sivaraj N, Sultan SM and Datta M 2016. Diversity analysis of grain characteristics and micronutrient content in rice landraces of Tripura, India. *Bangladesh. J Bot* **45**: 1143-49.

Dixit S, Singh UM, Abbai R, Ram T, Singh VK, Paul A, Virk PS, Kumar A 2019. Identification of genomic region(s) responsible for high iron and zinc content in rice. *Sci Rep* **9**:8136. https:// doi.org/10.1038/s41598-019- 43888- y

Du J, Zeng D, Wang B, Qian Q, Zheng S, Ling HQ 2013. Environmental effects on mineral accumulation in rice grains and identification of ecological specific QTLs. *Environ Geochem Health*, **35**: 161-170.

Franz KB, Kennedy BM, Fellers DA. 1980. Relative bioavailability of zinc from selected cereals and legumes using rat growth. *J Nutr.* 110:2272– 83. doi: 10.1093/jn/110.11.2272

Gande NK, Kundur PJ, Soman R, Ambati R, Ashwathanarayana R, Bekele BD, Shashidhar HE (2014) Identification of putative candidate gene markers for grain zinc content using recombinant inbred lines (RIL) population of IRRI38 × Jeerigesanna. *Afr J Biotechnol* **13**:657–663

Garcia-Oliveira, A. L., Tan, L., Fu, Y. and Sun, C., 2009. Genetic identification of quantitative trait loci for contents of mineral nutrients in rice grain. *J Integr Plant Biol.* **51**, 84.

Goffinet, B. & Gerber, S. 2000. Quantitative trait loci: a meta-analysis. *Genetics* **155**, 463–73.

Gregorio GB, Senadhira D, Htut H, Graham RD. 2000. Breeding for trace mineral density in rice. *Food Nutr Bull.* **21**:382– 6. doi: 10.1177/156482650002100407

Gregorio GB. 2002. Progress in breeding for trace minerals in staple crops. Symposium: plant breeding: a new tool for fighting micronutrient malnutrition. *J. Nutr.* **132**:500S–2S. doi: 10.1093/jn/132.3.500S

He, X. Y., Singh, P. K., Schlang, N., Duveiller, E., Dreisigacker, S., Payne, T., *et al*. 2014. Characterization of Chinese wheat germplasm for resistance to Fusarium head blight at CIMMYT, Mexico. *Euphytica* **195**, 383–395. doi: 10. 1007/s10681-013-1002-3

Huang Y, Tong C, Xu F, Chen Y, Zhang C, Bao J. 2016. Variation in mineral elements in grains of 20 brown rice accessions in two environments. *Food Chem.* **192**:873–8.doi: 10.1016/j.foodchem.2015. 07.087

Hu BL, Huang DR, Xiao YQ , Fan YY, Chen DZ, Zhuang JY. 2016. Mapping QTLs for mineral element contents in brown and milled rice using an *Oryza sativa* × *Oryza rufipogon* backcross inbred line population. *Cereal Research Communications*. **44**:57

Impa SM, Gramlich A, Tandy S, Schulin R, Frossard E, Johnson Beebout SE 2013. Internal Zn allocation influences Zn deficiency tolerance and grain Zn loading in rice (*Oryzasativa* L.). *Front Plant Sc* **4**:534

Indurkar AB, Majgahe SK, Sahu VK, Vishwakarma A, Premi V, Shrivastatva P. 2015. Identification characterization and mapping of QTLs related to grain Fe, Zn and protein contents in Rice (*Oryza sativa* L.). *Electronic Journal of Plant Breeding*. **6**:1059.

Ishikawa R, Iwata M, Taniko K, Monden G, Miyazaki N, Orn C, *et al*. 2017. Detection of quantitative trait loci controlling grain zinc concentration using Australian wild rice (*Oryza meridionalis*) a potential genetic resource for biofortification of rice. *PLoS One.* 2017;**12**:e0187224

Iron-related dietary pattern increases the risk of poor cognition. Nutrition journal 18:48

Jeomho L, Kyuseong L, Hunggoo H, Changihn Y, Sangbok L, Younghuran C, *et al.* 2008. Evaluation of iron and zinc content in rice germplasm. *Korean J Breed Sci.* **40:**101–5.

Jeng TL, Lin YW, Wang CS, Sung JM. 2012. Comparison and selection of rice mutents with high and zinc contents in their polished grains that were mutated from the indicatype cultiver IR64. *J Food Compost Anal.* **28**:149–54. doi: 10.1016/j.jfca.2012.08.008

Jin T, Chen J, Zhu L, Zhao Y, Guo J, Huang Y 2015. Comparative mapping combined with homology- based cloning of the rice genome reveals candidate genes for grain zinc and iron concentration in maize. *BMC Genet,* **16**(1):17

Joshi AK, Crossa J, Arun B, Chand R, Trethowan R, Vargas M, Monasterio IO (2010) Genotype × environment interaction for zinc and iron concentration of wheat grain in eastern Gangetic plains of India. *Field Crops Res*., **116**:268. https://doi.org/10.1016/j.fcr.2010.01.004

Kumar J, Jain S, Jain RK. 2014. Linkage mapping for grain iron and zinc content in F2 population derived from the cross between PAU 201 and Palman 579 in rice (*Oryza sativa* L.). *Cereal Research Communications.* **42**:389.

Kumar N, Jain RK, Chowdhury VK. Linkage 2019. Mapping of QTLs for Grain Minerals (Iron and Zinc) and Physio- Morphological Traits for Development of Mineral Rich Rice (*Oryza sativa* L.) *Indian Journal of Biotechnology*. **18**:69-80

Maganti S, Swaminathan R, Parida. 2020. A Variation in Iron and Zinc Content in Traditional Rice Genotypes. *Agric Res*, **9**:316–328.

Mishra A, Nanda SK 2008. Soils of Deogarh District. *Orissa Review* 44-46

Nachimuthu VV, Robin S, Sudhakar D, Raveendran M, Rajeswari S, Manonmani S. 2014. Evaluation of rice genetic diversity and variability in a population panel by principal component analysis. *Indian J Sci Technol*. **7**:1555–62.

Naik SM, Raman AK, Nagamallika M, Venkateshwarlu C, Singh SP, Kumar S, Singh SK, Ahmed HU, Das SP, Prasad K, Izhar T, Mandal NP, Singh NK, Yadav S, Reinke R, Swamy BPM, Virk P, Kumar A. 2020. Genotype × environment interactions for grain iron and zinc content in rice. *J Sci Food Agric* **100**:4150. https://doi.org/10.1002/jsfa.10454

Norton GJ, Deacon CM, Xiong L, Huang S, Meharg AA, Price AH. 2010. Genetic mapping of the rice genome in leaves and grain: Identification of QTLs for 17 elements including arsenic, cadmium, iron and selenium. *Plant and Soil.* **329**:139

Lee S-M, Kang JW, Lee JY, Seo J, Shin D, Cho JH, Jo S, Song YC, Park DS, Ko JM, Koh HJ, Lee J H 2020. QTL analysis for Fe and Zn concentrations in rice grains using a doubled haploid population derived from a cross between rice (*Oryza sativa*) cultivar 93-11 and milyang 352. *Plant Breed Biotech* **8**(1):69–76.

Lu K, Li L, Zheng X, Zhang Z, Mou T, Hu Z. 2008. Quantitative trait loci controlling Cu, Ca, Zn, Mn and Fe content in rice grains. *Journal of Genetics*. **87**:305 *Agronomy* **18**

Paltridge N, Palmer L, Milham P, Stangoulis J 2012. Energy-dispersive x-ray fluorescence analysis of zinc and iron concentration in rice and pearl millet grain. *Plant and Soil* **361**:251–260

Pandian SS, Robin S, Vinod KK, Rajeswari S, Manonmani S, Subramanian KS, Saraswathi R, Kirubhakaran APM 2011. Influence of intrinsic soil factors on genotype-by-environment interactions governing micronutrient content of milled rice grains. *AJCS* **5**(13):1737–1744

Pusadee T, Jamjod S, Chiang YC, Rerkasem B, Schaal BA 2009. Genetic structure and isolation by distance in a landrace of Thai rice. *Proc Nat Acad Sci* USA **106**:13880–5

Rao SP, Madhubabu P, Kota S, Bhadana VP, Varaprasad GS, Surekha K, Neerja CN, Babu VR 2014. Assessment of iron and zinc variability in rice germplasm using energy dispersive x-ray florescence spectrophotometer (ED-XRF). *J Rice Res* 7:45–52

Raza, Q., Riaz, A., Sabar, M., Atif, R. M., and Bashir, K. 2019. Meta-analysis of grain iron and zinc associated QTLs identified hotspot chromosomal regions and positional candidate genes for breeding biofortified rice. *Plant Sci.* **288**:110214. doi: 10.1016/j.plantsci.2019.110214

Reinike R, Slamet-Loedin IH 2016. Advances in breeding for high grain Zinc in rice. Rice, **9**:49. https://doi.org/10.1186/s12284- 016- 0122- 5.

Roy SC, Sharma BD 2014. Assessment of genetic diversity in rice (*Oryza sativa* L.) germplasm based on agro-morphology traits and zinc-iron content for crop improvement. *Physiol Mol Biol Plants* **20**:209–224.

Sanghamitra P, Bose LK, Bagchi TB *et al.* 2022. Characterization and exploring genetic potential of landraces from Odisha with special emphasis on grain micronutrient content for benefaction of biofortification in rice. *PhysiolMolBiol Plants,* **28**: 203–221,https://doi.org/10.1007/s12298-021-01119-7

Sahu GC, Mishra A 2005. Soil of Orissa and Its Management Orissa Review: 56-60

Saleh AS, Wang P, Wang N, Yang L, Xiao Z 2019. Brown Rice Versus White Rice: Nutritional Quality, Potential Health Benefits, Development of Food Products, and Preservation Technologies. *Compre Rev Food Sci Food Safety* **18**:1070–96

Shi Z, El-Obeid T, Li M, Xu X, Liu J (2019) Iron-related dietary pattern increases the risk of poor cognition. *Nutrition Journal* **18**:48.

Stangoulis, J. C., Huynh, B. L., Welch, R. M., Choi, E. Y. and Graham, R. D., 2007. Quantitative trait loci for phytate in rice grain and their relationship with grain micronutrient content. *Euphytica.* **154**, 289.

Stangoulis J, Sison C 2008. Crop sampling protocols for micronutrient analysis. Harvest Plus Tech MonogrSer 7. International FOOD Policy Research Institute (IFPRI) and International Centre for Tropical Agriculture (CIAT), Washington DC, Cali

Suman K, Neeraja CN, Madhubabu P, Rathod S, Bej S, Jadhav KP, Kumar JA, Chaitanya U, Pawar SC, Rani SH, Subbarao LV and Voleti SR (2021) Identification of Promising RILs for High Grain Zinc Through Genotype x Environment Analysis and Stable Grain Zinc QTL Using SSRs and SNPs in Rice (*Oryza sativa* L.). *Front. Plant Sci.* **12:**587482. doi: 10.3389/fpls.2021.587482

Soe YP (2020) Meta-analysis of quantitative trait loci associated with grain zinc content in rice.
Dissertation. University of the Philippines, Los Baños. 180 pp

Sellappan K, Datta K, Parkhi V, Datta SK. 2009. Rice caryopsis structure in relation to distribution of micronutrients (iron, zinc, b-carotene) of rice cultivars including transgenic indica rice. *Plant Sci.* **177:**557– 62. doi: 10.1016/j.plantsci.2009.07.004

Swamy BM, Vikram P, Dixit S, Ahmed HU, Kumar A (2011) Meta-analysis of grain yield QTL identified during agricultural drought in grasses showed consensus. BMC Genomics 12:319. https://doi.org/10.1186/1471- 2164- 12- 319

Swamy BPM, Rahman MA, Inabangan-Asilo MA, Amprado A, Manito C, Chada-Mohanty P, Reinike R, Slamet-Loedin IH 2016. Advances in breeding for high grain Zinc in Rice. *Rice* **9:**49. https://doi.org/10.1186/s12284- 016- 0122- 5

Swamy BPM, Kaladhar K, Anuradha K, Batchu AK, Longvah T, Sarla N (2018a) QTL analysis for grain iron and zinc concentrations in two *O. nivara* derived backcross populations. *Rice Sci* **25**(4):197–207. https://doi.org/10.1016/j.rsci.2018.06.003

Swamy BPM, Descalsota GIL, Nha CT, Amparado A, Inabangan-Asilo MA, Manito C, Tesoro F, Reinke R (2018b) Identification of genomic regions associated with agronomic and biofortification traits in DH populations of rice. *PLoS One* **13**(8):1–20. https://doi.org/10.1371/journal. pone.0201756

Thongbam PD, Raychaudhury M, Durai A, Das SP, Ramesh T, Ramya KT *et al.* Studies on grain and food quality traits of some indigenous rice cultivars of north-eastern hill region of India. *J Agric Sci.* (2011) 4:259– 70. doi: 10.5539/jas.v4n3p259

Tripathy SK (2020) Genetic variation for micronutrients and study of genetic diversity in diverse germplasm of rice. *J Crop and Weed* **16**:101-109

Vanlalsanga S, Singh P , Tunginba Y Singh 2019. Rice of Northeast India harbor rich genetic diversity as measured by SSR markers and Zn/Fe content. *BMC Genetics* **20**:79

Wang Y, Wei Y, Dong L, Lu L, Feng Y, Zhang J *et al* (2014) Improved yield and Zn accumulation for rice grain by Zn fertilization and optimized water management. J Zhejiang Univ Sci B **15**(4):365–374. https://doi.org/10.1631/jzus.b1300263

White PJ, Broadley MR 2009. Biofortification of crops with seven mineral elements often lacking in human diets – iron, zinc, copper, calcium, magnesium, selenium and iodine. Review. *New Phytol* **182**:49–84. https://doi.org/10.1111/j.1469-8137.2008.02738. x

Wissuwa M, Ismail AM, Graham RD (2008) Rice grain zinc concentrations as affected by genotype, native soil-zinc availability, and zinc fertilization. *Plant Soil* **306:**37. https://doi.org/10.1007/s11104-007- 9368- 4

Wailes E, Chavez E. International rice baseline with deterministic and stochastic projections, 2012-2021. Staff Pap. (2012)

Wessells KR, Brown KH. 2012. Estimating the global prevalence of zinc deficiency: results based on zinc availability in national food supplies and the prevalence of stunting. *PLoS ONE.* 7:e50568. doi: 10.1371/Journal.Pone.0050568

Welch RM, Graham RD. 2004. Breeding for micronutrients in staple food crops from a human nutrition perspective. *J Exp Bot.* 55:353– 64. doi: 10.1093/jxb/erh064

Wynne JC and Gregory WC 1981. Peanut Breeding. *Advances in Agronomy* **34**: 39–71.

Vasconcelos M, Datta K, Oliva N, Khalekuzzaman M, Torrizo L, Krishnan S, Oliveira M, Goto F, Datta K 2003. Enhanced iron and zinc accumulation in transgenic rice with the ferritin gene. *Plant Sci* **164**:371–378.

Vanlalsanga S, Singh P , Tunginba Y Singh 2019. Rice of Northeast India harbor rich genetic diversity as measured by SSR markers and Zn/Fe content. *BMC Genetics* **20**:79.

Varshney, R. K., Singh, V. K., Kumar, A., Powell, W., and Sorrells, M. E. (2018). Can genomics deliver climate-change ready crops? *Curr. Opin. Plant Biol.* **45**(Pt B), 205–211. doi: 10.1016/j.pbi.2018.03.007

Xu Q , Zheng TQ , Hu X, Cheng LR, Xu JL, Shi YM, *et al.* 2015. Examining two sets of introgression lines in rice (*Oryza sativa* L.) reveals favorable alleles that improve grain Zn and Fe concentrations. *PLoS One.* **10**:e0131846

Zaw H, Raghavan C, Pocsedio A, Swamy BPM, Jubay ML, Singh RK *et al* 2019. Exploring genetic architecture of grain yield and quality traits in a 16-way indica by japonica rice MAGIC global population. *Sci Rep* **9**(1):1–11. https://doi.org/10.1038/s41598- 019-55357- 7

Zhang X, Zhang G, Guo L, Wang H, Zeng D, Dong G, *et al.* 2011. Identification of quantitative trait loci for Cd and Zn concentrations of brown rice grown in Cd-polluted soils. *Euphytica.* **180**:173

Zazoli MA, Shokrzadeh M, Bazerafshan E, Hazrati M, Tavakkouli A. 2006. Investigation of zinc content in Iranian rice (*Oryza sativa*) and its weekly intake. *Am Eurasian J Agric Environ Sci.* **1:**156–9.

6

Other Nutritional and Anti-Nutritional Factors: Role and Scope in Biofortification

Introduction

Rice is the world's most important staple crop and is a primary source of energy and nutrition for about half of the world's population. So it is the most important cereal crop with respect to food security of common people. Due to increasing world population and shrinkage of farm land holdings there is an urgent need to increase the grain productivity. Therefore, increasing grain yield has become the primary objective in many rice-breeding programs. India is the second highest rice producer (about 120 million ton/year) in the world and able to meet the demand of rice for the entire Indian population. But along with yield improvement of rice, quality has now become a foremost consideration for rice consumers and millers. Quality is defined as "the totality of features and characteristics of a product or service that bears its ability to satisfy stated or implied needs" (International Standard Organization (ISO) 8402 1986). The concept of grain quality covers many physico-chemical properties such as grain shape and size, milling and cooking properties, amylose content, stickiness, chalkiness, texture, colour, gelatinization characters and various nutritional properties. Thus, rice grain quality generally includes five main classes i.e. (i) milling quality, (ii) physical quality, (iii) cooking and eating quality (iv) sensory quality and (v) nutritional quality. Grain quality and its assessment are very important to consumers, millers and rice breeders who are engaged in varietal improvement programme, related to new features such as high nutritional quality, high yield potential, highly resistant to abiotic or biotic stresses, higher water use efficiency etc.

Now a days, fortification in food is a trend amongst the people of developed nations. Fortification is the process of adding essential micronutrients, minerals and other health-promoting compounds to foods through any external source. But in developing countries, it is very unsustainable programme due to lack of execution, suitable technology and consumer awareness. In contrast, biofortification is a more sustainable strategy because this involves the fortification of crops especially cereals and legumes at the source. It is the inherent capability of the plants to accumulate more nutritional components in their edible parts. Biofortified rice includes the grain which is enriched with nutritional components in higher quantity as compared to average content. Two types of this rice is available: (i) naturally found germplasm /cultivars, in which some nutritional compounds are present in higher amount than average. (ii) nutritionally enriched rice varieties developed by the scientist through conventional breeding, biotechnological approach or nutrient management. The naturally biofortified rice have many demerits with respect to yield, plant types, physico-chemical quality of the grain, environmental factors of cultivation etc. Therefore, this type of rice is less popular amongst the farmers and millers though it contains higher nutritional components. The scientists always try to utilize this naturally evolved biofortified rice in their breeding programme to minimize the demerits and develop a new variety with many desirable characteristics.

What should be the quality of a biofortified rice? Rice is utilized in different forms according to the consumer's preference. After harvesting, the raw grains are processed in different ways to make brown rice, polished raw rice, polished parboiled rice, unpolished parboiled rice, popped rice, puffed rice, flaked rice, boiled rice, raw rice flour etc. The nutritional components may differ in accordance with the grain processing methods. As for example, brown rice is more nutritious than the polished raw rice due to the presence of bran layers attached with endosperm. In a whole rough rice grain the composition of different parts are Hull -20%, bran & germ-10% and endosperm is 70% (Orthoefer, 2005).Rice bran, a major co-product in the rice milling industry, usually contains about 11.3–14.9% protein, 34.0–62.0% carbohydrates (mainly starch), and 15.0–19.7% oil (Juliano, 1993).It is also a good source of vitamin E, oryzanol components (ferulate esters of triterpene alcohols) (Nicolosi, Rogers, Ausmann, & Orthoefer, 1993), Ca, Mg, Fe, Zn, diatery fiber (20.3% DM) (Phimolsiripol *et al.*, 2012), phenolic acids, flavonoids, anthocyanin and phytic acid (Iqbal, Bhanger, & Anwar, 2005). These vitamins and antioxidative compounds shows antitumor properties in mammary cancer (Nasaretnam, 2005), hypo-cholesterolmic and anti-inflammatory (Akihisa *et al.*, 2000) effects.

Therefore, rice bran is a very important component of the grains. But during milling, the bran and germs are entirely or partially removed from the endosperm to get white rice, which is generally consumed. During parboiling of the grain, some lipophilic nutritional components viz. tocopherol, tocotrianol migrate from the bran layers towards the endosperm to avoid water molecules. On the other hand some hydrophilic nutritional components viz. phenolic acid, flavonoids etc. become washed out. But some researcher reported that the parboiling and storage of grain affected vitamin E content with average loss of 60% as compared to nonparboiled rice. Among these vitamin-E homologues, α-tocopherol was very sensitive and 93% loss is found after parboiling as compared to original content, whereas γ-tocotrienol content was affected the least (Pascual *et al*., 2013). Thus, different hydrothermal processing of the grain alters the content of nutritional components in processed products. Therefore, post harvest processing of the grain in very important with respect to bio-fortification of rice.

Bio-fortified rice should contain one or more bio-available nutritional compounds in higher amount and minimum level of anti-nutritional factors as compared to average quantity and having comparable yields, grain quality and plant types with the most popular rice genotypes.

Some Nutritional Components of Rice Grain, Having Antioxidant and Medicinal Properties

Milled rice consists of about 60 to 78% starch, while protein represents the second most abundant constituent (6–7%). Conversely, rice bran contains high levels of fat (15–20%), minerals and proteins (11–15%), simultaneously, it is a good source of fiber (7–11%). The moisture content of mature grain at harvesting stage should be 13 to 14% but after sun drying the grains are generally stored at 12 to 13% moisture content. Rice bran displays much higher levels of minerals and vitamins than the other rice milling fractions (Juliano, 1985). However, the whole rice grain contains different antioxidative compounds in very trace amount other than carbohydrates, proteins, fats, fibers, moisture and minerals. The brief description of these compounds are given below.

Phenolic acids

Prevoius study reported that the pigmented rice samples had higher content of phenolic compounds than non-pigmented samples and the pericarp layers (bran) was the major source of various phenolic acids (ferulic acid, *p*-coumaric acid, vanillic acid, *p*-hydroxybenzoic, gallic acid, and protocatechuic acid) (Butsat and Siriamornpun 2010). In addition, the brown rice was also rich in different phenolic compounds (0.05 to 7.69 mg/100g) than the polished rice

(0.02 to 1.12mg/100g) with the predominance of p-coumaric acid, ferulic acid, gallic acid and protocatechuic acid (Vichapong *et al*., 2010).The free phenolics content in black rice bran ranged from 2086 to 7043 mg of GAE/100 g (DW) constituting 88.2 to 95.6% of total acids. The bound phenolic content ranged from 221.2 to 382.7 mg of gallic acid equiv/100 g (DW) (Zhang *et al*., 2010). In fact, Phenolic compounds exist in three forms: the free, soluble conjugate (esterified), and insoluble-bound forms (Jung *et al*., 2002). Due to presence of carboxylic acid and hydroxyl groups in their structures, phenolic acids form both ester and ether bonds with other compounds resulting the formation of linkages with cell wall polysaccharides (Yu *et al*., 2001). Phenolic acids can be esterified with small molecules like alcohols, other phenolic acids and alkaloids to form soluble conjugated structure. Bound phenolic acids are typically involved in cell wall structure (Bunzel *et al*., 2002; Sun, Sun, & Zhang, 2001, 2002) where cross-linking of lignin components via phenolic acids appears to have a profound effect on the growth of the cell wall and its mechanical properties and biodegradability. Researcher also demonstrated the presence of anthocyanins in the bran layer of pigmented rice, particularly, cyanidin 3-glucoside and peonidin 3-glucoside. The anthocyanins also contributed to marked antioxidant activities in vivo by preventing DNA damage and LDL deterioration. However, Tang *et al*., 2015 reported that there was 20 to 26% reduction of total phenolics content and 13 to 20 % reduction in total flavonoids content in some black rice during 25 minutes domestic cooking. Salgado *et al*., 2010 reported that diet containing black rice reduced the level of plasma cholesterol, triglycerides, and low-density lipoprotein and increase the level of good lipid (HDL) in rats.

Phenolic acids, flavonoids and anthocyanin are the main antioxidative compounds in pigmented rice. But there are other nutritional compounds or elements which are responsible for providing us the antioxidant activity as well as health benefits, present in pigmented rice in higher quantity as compared to white rice.

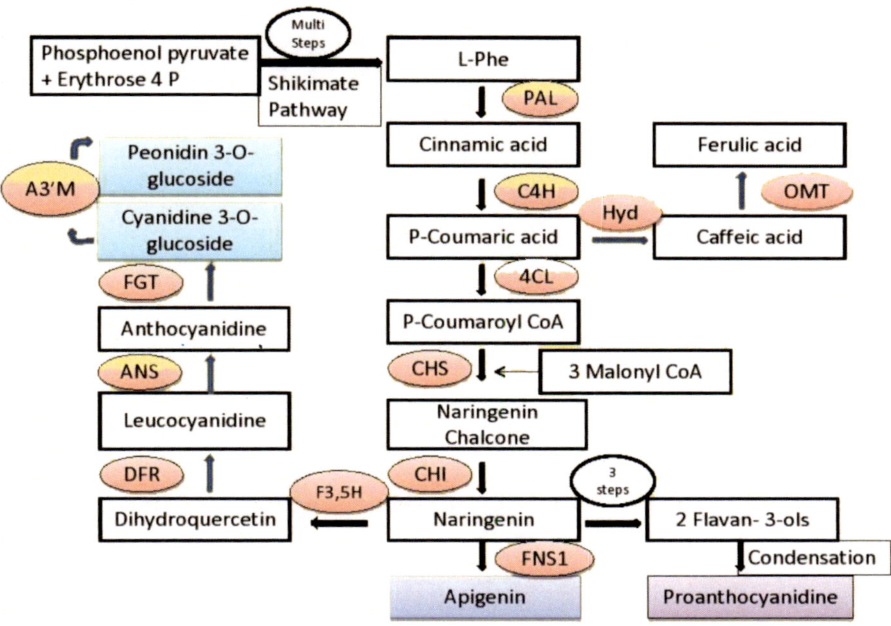

Fig. 1: Schematic simplified diagram of Phenolic acids, anthocyanin and proanthocyanidine synthesis in rice. Enzymes: PAL- Phenylalanine ammonia lyase; C4H-Cinnamate- 4-hydroxylase; 4CL- Coumarate CoA Ligase; CHS-Chalcone synthase; CHI-Chalcone isomerase. FNS1:Flavone synthase-1; F3,5H: Flavone 3'5'Hydroxylase; DFR: Dihydroflavonol 4 reductase; ANS: Anthocyanidine synthase; FGT:UDP-glucose flavonoid 3-O-glucosyl transferase; A3'M: S adenosyl methionine transferase.

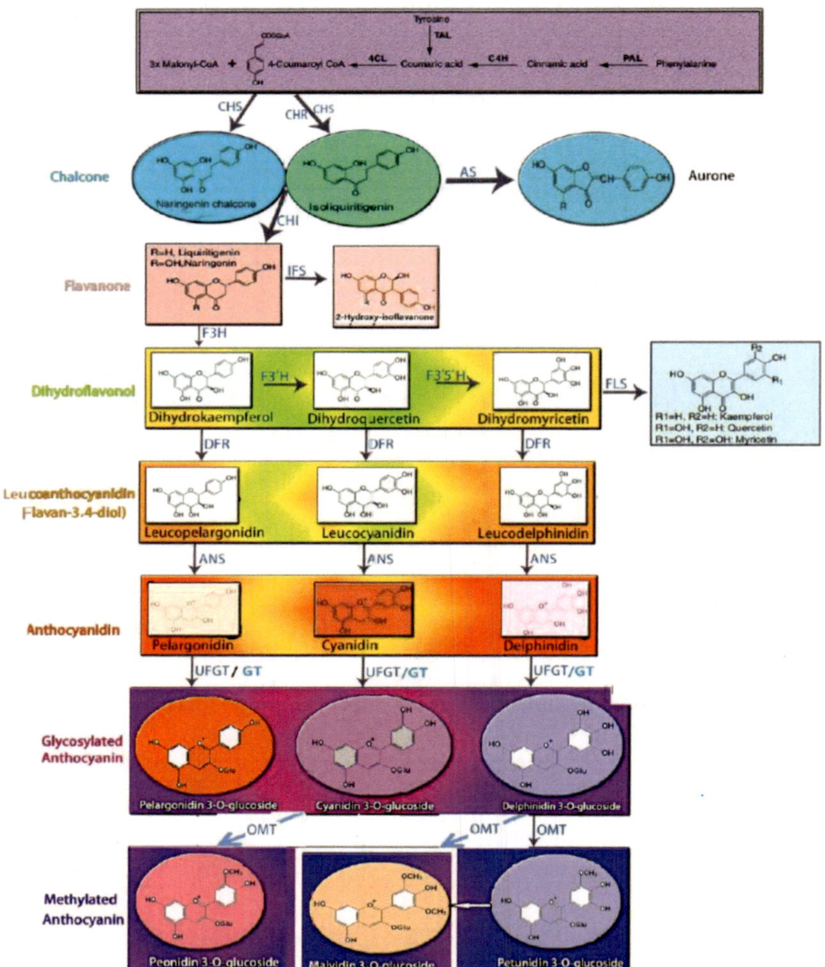

Fig. 2: Anthocyanin biosynthesis pathways in rice (Partly reproduced from Buchanan *et al.*, 2015, Biochemistry and molecular biology of Plants, American Society of Plant Biologist. pp1193).

γ-oryzanol

Chemically, γ-oryzanol is the esters of ferulic acid with phytosterols and triterpene alcohols, generally referred as steryl ferulates. Its' biosynthesis initiates with acetyl-CoA and form isopentenyl diphosphate (IPP) via mevalonic acid through some enzymatic steps. HPLC-PDA chromatogram of γ-oryzanol showed four main peaks in 325 nm at retention times of 19.4 (peak 1; cycloartenyl ferulate), 21.5 (peak 2; 24-methylenecycloartanyl), 23.5 (peak 3; campesteryl ferulate) and 26.8 (peak 4; sitosteryl ferulate) min (Cho *et al.*, 2012).

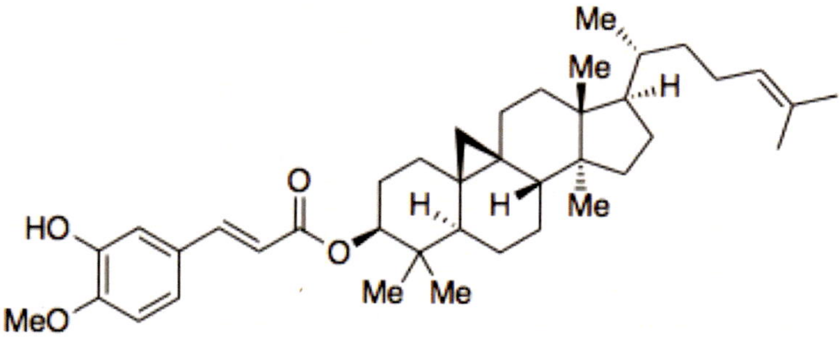

Fig. 3: Chemical structure of Gamma oryzanol

Previous study reported that γ-oryzanol content of 196 Korean landraces ranged from 9.8 to 55.9 mg 100 g^{-1} and a mean was 27.2 mg /100 g (Song *et al.*,2013). Bergman and Xu (2003) indicated that genotype and environment had a significant role in controlling gamma oryzanol level per grain weight. However, cooking generally decrease the γ-oryzanol content of the grain with a few exception. Similarly, parboiling results in an average γ-oryzanol loss of 20% (Pascual *et al.*, 2013). γ- Oryzanol content of pigmented rice was higher (44.73mg/ml rice extract) than non pigmented (0.0 to 2.5 mg/ml rice extract) genotypes (Chakuton *et al.*, 2012). Banterng *et al.* (2014) also assessed 25 Korean black rice and reported that it contains higher amount (0.223 to 0.318 mg/g grain) of γ-oryzanol than white rice (0.168 to 0.214 mg/g grain). The concentration of γ-oryzanol varied in different components of the grain. The highest concentration of it was found in the rice bran, obtained from the first 10 s milling. In other words, it can be expressed that the highest γ-oryzanol concentration was found in the outer portion of the bran layer (Ha *et al.*, 2006). γ-Oryzanol plays a significant role in various biological activities, including anti-inflammatory (Akihisa *et al.*, 2000), antioxidant (Ismail *et al.*, 2010), and anti-tumor activities (Yasukawa *et al.*, 1998). In addition, γ-oryzanol has also been reported to lower blood cholesterol levels in humans (Wilson *et al.*, 2007).

Vitamin-E

Vitamin E was discovered by Even and Bishop. Vitamin E is indispensable for reproduction and prevents diseases associated with oxidative stress, such as cardiovascular disease, cancer, chronic inflammation and neurologic disorders. Vitamin E has an additional positive effect on autoimmune disease by decreasing proinflammatory cytokines, lipid mediators and has important role in rheumatoid arthritis patients. Fifty to 70% of children with chronic cholestasis are vitamin E deficient, even when consuming standard supplements of vitamin E. α-tocopherol is the most bioactive form of vitamin E followed in order by β, γ and δ tocopherol. The most commonly used pharmacologic form of α-tocopherol is a totally synthetic product and has both D and L stereoisomers now designated as racemic α-tocopherol.

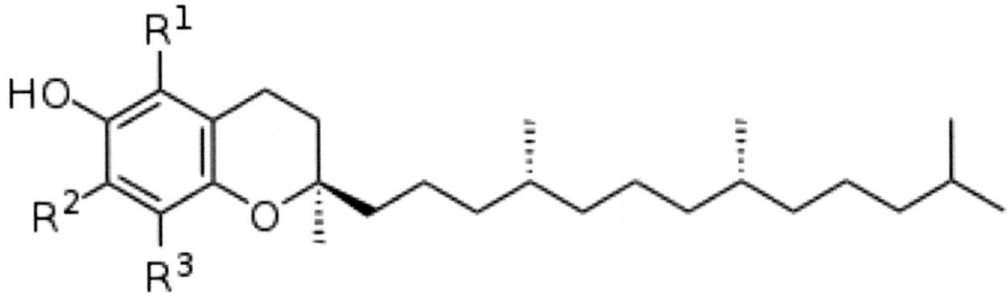

Fig. 4. Chemical structure of Tocopherol (Vit. E)

This synthetic form has considerably less bioactivity (approximately 75%) than the pure naturally occurring form. Vitamin E is absorbed into intestinal mucosa by nonsaturable, noncarrier-mediated passive diffusion process and absorption is maximal in median small intestine and none is absorbed in large intestine. Human body stores 40 mg/kg vitamin E, 77% of which is in adipose tissue, 20% in muscle and only 1% in liver. Vitamin E may be beneficial in preventing prostate cancer or delaying disease progression.Eight forms of tocochromanols are commonly considered to occur in plants, i.e. four forms each of tocopherol and tocotrienol, carrying a phytyl or geranylgeranyl side chain respectively.Tocopherols are synthesized from homogentisate and phytyl-diphosphate by homogentisate phytyltransferase (HPT1).In rice, vitamin E homologs are varied according to the genotypes and post harvest processing. In Brazil, thirty-two different samples of dehulled brown rice (Indica and japonica) were analysed for tocopherol and tocotrianol content and observed that total vitamin E levels varied largely at 10.4–32.5 mg/kg. In, *japonica* subspecies, the average concentrations of different tocols from high to low were α-Tocopherol (10.0 mg/kg), α-Tocotrianol (7.0 mg/kg), γ- Tocotrianol (5.8 mg/kg), and γ- Tocopherol (1.4 mg/kg). In *indica* rice, the most abundant homolog was γ- Tocotrianol (7.8 mg/kg), followed by α-Tocopherol (4.8 mg/kg), then α- Tocotrianol (2.3 mg/kg), and γ-Tocopherol (1.3 mg/kg) (Heinemann *et al.*, 2008). Rice parboiling is a hydrothermal treatment consisting of soaking (at a temperature above 58 °C), autoclaving (120 °C), causing total or partial starch gelatinization, and drying. Khatoon and Gopalakrishna (2004) found that vitamin E levels were greatly reduced after parboiling. Parboiling affected mostly vitamin E, which presented an average loss of approximately 60% when compared to the initial level. Pascual *et al* 2013 also reported that cooking caused a further loss of tocols and the final amounts were irrelevant in both cooked non-parboiled and parboiled rice.

Other Vitamins

Brown rice contains vitamins like thiamin (B1), riboflavin (B2), niacin (B3), pantothenic acid (B5), pyridoxine (B6), biotin (B7), folate (B9), and a-tocopherol (E), but does not have vitamins A, D, or C (Juliano and Bechtel, 1985). But after milling the vitamins content is reduced bythiamin (B1) 68% - 82%, riboflavin (B2) by 57%, niacin (B3) by 64% -79%, pantothenic acid (B5) by 51% - 67%, B6 by 43% - 86%, folic acid (B9) by 60% - 67%, biotin (B7) 86%, vitamin E by 82%, fats by 77% - 82%, proteins by 10% - 16%, and fibers by 63% - 78% (Dexter, 1998). Bio-fortified rice with vitamin A through genetic manipulation is termed as golden rice. β-carotene is most prevalent form of vitamin A and it is the precursor of vitamin A. Phytoene synthase (PSY) is the important rate-limiting and regulatory enzyme in the carotenoid biosynthesis pathway. β carotene is produced from geranylgeranyl diphosphate with the help of some enzymes like PSY, Phytone desaturase, carotene desaturase, lycopene β-cyclase etc. These enzymes are absent in rice and thus rice cannot synthesize β carotene. PSY and carotene desaturase genes (from daffodil) were transformed into rice nuclear genome and placed under the control of an endosperm specific promoters for expression into the endosperm. The end product of this engineered pathway will be lycopene (red colour) which is further processed by endogenous enzymes to form β-carotene. Original golden rice contain 1.6μg/g carotene in endosperm.

Ghosh S. *et al.* (2019) reported that the vitamin B1 (Thiamin) present in brown rice is about 2.9- 6.1 mg/g, whereas the RDI value of vitamin B1 for the adults and children older than 4 years is approximately 1.2 mg per day. Thiamin pyrophosphate acts as a cofactor for the enzymes of basic metabolic pathways like glycolysis, pentose phosphate pathway, and TCA pathway. It is also involved in amino acid and acetyl Co-A biosynthesis pathways. In rice it is mainly found in bran and germ layers.Thiamin is also involved in biotic and abiotic stress response in plants.

Vitamin B2 or riboflavin is the water-soluble vitamin and the precursor of flavin adenine dinucleotide (FAD) and flavin mononucleotide (FMN), which act as cofactors in different metabolic enzymes. These cofactors are involved in mitochondrial electron transport, fatty acid synthesis, and in the metabolism of vitamins B6, B12, and folates. In brown rice, it is present in minute quantity i.e.0.4-1.4 µg/g.

Vitamin B3 (niacin), is a water-soluble vitamin that can exist as nicotinic acid or nicotinamide. It acts as a cofactor of NAD and NADP. These are the key cofactors involved in carbohydrate, protein and fat metabolism. In brown rice it is also present (32-53 µg/g) in bran layers.

Vitamin B5, or pantothenate or pantothenic acid, is a water-soluble vitamin that is synthesized by microorganisms and plants. Pantothenate is the precursor of the 4' -phosphopantetheine moiety of CoA and acyl-carrier protein, which is involved in fatty acid metabolism and lignin biosynthesis. In brown rice it is also present in minute quantity (9-15 µg/g) in bran layers. The vitamin B6 (Pyridoxine, pyridoxamine, and pyridoxal) content of brown rice is 5-9 mg/g, which is reduced to 43% - 86% after milling. There is a scope to enhance vitamin B6 content in rice grain through overexpression of Arabidopsis PDX1 and PDX2 genes under CAMV35S promoter system which increased vitamin B6 content of cassava in the leaves and roots respectively (Li *et al.*, 2015).

Folate Biofortification

Folates (vitamin B9) jointly refer to a group of structurally related folate derivatives. In contrast to the lipid-soluble micronutrient vitamin A, folates are water-soluble. Folates act as cofactors in one-carbon metabolism, required for amino acid and nucleotide metabolism (Hansonand Gregory, 2011). Notably, folates work a critical role in neurotransmitter formation and neural tube growth in humans (Imbard *et al.*, 2013). The dietary requirement (RDAs) for folates are 400 mg/day for adults and 600 mg/day for pregnant women (WHO, 2004). Plants serve the chief dietary sources of folates.

To establish the natural variations of folates in cultivated rice, the folate content was estimated in 78 rice varieties and about eight-fold difference in folates was found in both brown and milled grains (Dong *et al.*, 2011). The folate levels was found generally lower in milled grains (up to 78 mg/100 g DW) as compared to whole grains (up to 111 mg/100 g DW) (Dong *et al.*, 2011). Another study of taking 12 rice varieties showed a two-fold variation in total grain folates (up to 70 mg/100 g DW) (Blancquaert *et al.*,2010). Major effect QTLs associated with high folate content in milled rice grains was identified in recombinant and backcross derived lines (Dong *et al.*, 2014). But, those QTLs could not be mapped to the chromosomal locations of folate biosynthetic genes (Dong *et al.*, 2014).

Glycemic Index of Rice and Human Health

The glycemic index (GI) is a scale that ranks carbohydrate-rich foods according to the raise of blood glucose levels during a certain period of time (Jenkins *et al.*,1981) to represent the physiological effect of carbohydrates in foods.The GI is defined as the incremental area under the blood glucose curve (IAUC) of a 50 g carbohydrate portion of a particular food expressed as a percentage of the response to 50 g carbohydrate of a standard food taken by the same subject, on a different day. The GI means how much each gram of available carbohydrate in a food raises blood glucose levels following consumption of the food, relative to consumption of pure glucose (Jenkins *et al.*,1981). The rates of digestion and absorption of glucose in the small intestine determine the GI of starchy foods. For nutritional purposes, starch can be classified into rapidly digestible starch (RDS), slowly digestible starch (SDS), and resistant starch (RS) according to *in vitro* digestion.RS in particular, being the indigestible starch portion in the stomach and digested in the small intestine, can help in body weight management, and in the prevention of heart disease, cancer, diabetes, and cardiovascular disease, as well as providing functional control of the GI. There are three ratings of GI for foods (Brand-Miller *et al.*, 2003) viz., low GI foods (GI, 55 or less), medium GI foods (GI of 56-69) and high GI foods (GI, 70 or more).The amylose content and GI of rice may be correlated exponentially rather than linear. Higher amylose rice have been shown to have slower rate of starch digestion and glycemic responses than rice varieties with low amylase content (Foster-Powell *et al.*, 2002). Besides, the presence of dietary fiber in rice affects glucose response and the GI negatively. The amylose content in combination with dietary fiber contributes significantly to lower down the glycemic response and the GI of rice. Many researchers reported that parboiling of rice grain decreased the glucose level of serum and thus lower the GI value. Consumption of brown rice also lowers the GI value at 12.1 % (for non diabatic people) to 35.6% (for diabatic people).The result was due to the higher content of phytic acid, dietary fiber, polyphenols, and lipid in brown rice as compared to white rice, and variation in some physicochemical properties of grain, viz. cooking time and degree of gelatinization. According to FAO and WHO (1998) GI grading, *kattuyanam* (brown rice) was graded as low GI rice (47.19), while *red* and *black kavuni* was graded as medium GI rice (61.69 and 56.27) and white rice *karudan samba* as high GI rice (69.74). Somaratne *et al.*, (2017) reported that red and white basmati rice have low GI but after polishing GI value increased marginally. Kumar *et al.* (2017) demonstrated that the variety Swarna and Mashuri showed mediun GI value and higher amount of resistant starch content (2.37 to 2.57%). According to NRRI, annual report (2013-14) improved Lalat (MAS) showed lower GI value (about 55) but exploration and biochemical study of more germplasm is required to get promising variety for diabatic people.

Anti-nutritional Factors Present in Rice and their Role

Anti-nutritional factors are the bio molecules which have detrimental effects to humans and animal growth by impairing uptake or utilization of other foods and feed components or generate discomfort and stress. It is mainly occurred in pulse but rice grains have some anti-nutritional factors mainly concentrated in bran layers viz. Phytic acid, alpha amylase inhibitor, trypsin inhibitor, polyphenols, toxic chemicals, heavy metals and arsenic. Phytic acid (PA), known as myo-inositol 1,2,3,4,5,6- hexa-kisphosphate, often exists in the form of a mixed salt (phytate or phytin) of mineral 2+ 2+ 3+ cations, including Zn , Ca and Fe. It reduces the bioavailability of minerals in our intestine due to formation of complex. Phytic acid digestive enzyme (Phytase) is absent in our intestine. It also reduces the activity of protein and starch digestive enzymes in our digestive tract. In raw rice grain the range of phytic acid varied between 0.1 to 3.0 % depending on genotypes. Phytic acid content is varied due to genotypes and environment but there was no correlation between protein content and phytic acid content in grain (Liu *et al.*, 2005).Rice bran, consisting of pericarp, aleurone and germ, has a high concentration of phytic acid ranging from 5.94 to 6.09 g/ 100 g (Kasim and Edwards, 1998). IPKI is the regulatory enzyme that play an important role in phytic acid biosynthesis. The rice IPKI gene (Os04g0661200) is highly expressed in developing grain and activation or suppression of this gene through RNAi gene silencing may result in alteration of the biosynthesis and accumulation in the grain (Kumar *et al.*, 2017). However, in some cases phytic acid acts as an antioxidants, anticancerous and hypoglycaemic agent but the threshold level of phytic acid content in rice grain for our consumption, is not established clearly. In some study, it was shown that phytic acid content of rice was reduced after different hydrothermal processing of the grains.

Alpha-amylase and trypsin inhibitor are the agents which impair the activity of digestive enzymes for carbohydrate and protein and it was observed that their content in rice is lower than wheat and pulses. Polyphenols are the compounds which generally act as antioxidants but sometimes in higher concentration, it is responsible for generating imbalance in proton gradient generated in mitochondrial electron transport chain resulting disturbance in ATP formation. Therefore, it may acts as an anti-nutritional factor. In rice, total phenol content ranged between 100 to 1500 mg GAE/100g depending on genotypes and environmental factors for rice cultivation. Heavy metals like Mercury, Cadmium, Chromium, and Lead are toxic metals and their accumulation in our body can cause serious effects like anaemia, hypertension and kidneys, lungs, bones problem. Arsenic is the metalloid and contaminates the ground water of intensive rice growing area. According to FAO, the allowable limit for Mercury, Cadmium, Lead and Arsenic in rice grain is 0.1, 0.1,0.2 and 0.2µg/g respectively.High As accumulation (mean = 143 ± 76 µg/kg; range = 2–557 µg/kg; n = 214) has been reported in rice grain cultivated through groundwater irrigation (Rahman *et al.*, 2009). In drinking water, the maximum contamination levels of arsenic were lowered from 50 µg/L to 10 µg/L by USEPA and WHO (IARC, 2004). Arsenic in drinking water is mainly inorganic arsenic and more harmful than in food such as grains and vegetables (Akter *et al.*, 2005). Inorganic arsenic is more toxic than organic arsenic and can exist in two oxidation states: arsenite (As3+), and arsenate (As5+). Mandal *et al.*, 2020 showed that arsenic concentration in cooked rice (n = 100) ranged from 31.5 to 80 µg/kg as compared to raw rice (105–510 µg/kg) when cooked with As safe water (< 0.5 µg/l). Approximately, 89.9% of the total As prevalent as inorganic species (arsenite and arsenate) and remaining as dimethyl arsinic acid in raw rice grain (Roychowdhury, 2008a). Cadmium (Cd), one of the most hazardous heavy metals, is readily absorbed by rice (*Oryza sativa L.*) and subsequently enters human body through food chain (Xie *et al.*, 2015). The mean Cd concentration of 484 rice samples obtained from different contaminated areas of China ranged from 149 to 189 µg/kg and indica rice accumulates more Cd than japonica in the grain (Ke at al.,2015). The mean Cd concentrations in rice grain were found highest in Bangladesh (99 µg/kg) and Sri Lanka (81 µg/kg), followed by India (78 µg/kg) (Meharg *et al.*, 2013). The highest Cd concentrations in grain were 1310 µg/kg, 800 µg/kg and 1000 µg/kg for Bangladesh, Sri Lanka and India, respectively (Meharg *et al.*, 2013). The half-life of cadmium in humans is 10–30 years and one of the main causes of Itai-itai disease. Chronic poisoning of cadmium leads to impair calcium balance, causing osteoporosis, rickets, and fractures (Ahn *et al.*, 2016). Cadmium can also interact with various functional proteins, functional groups or nucleic acids of human, resulting in damage of DNA and occurrence of cancer or deformity (Luevano and Damodaran, 2014). It also damage kidney and cause death after intake of Cd rich food for long period.The complex substances are formed by cadmium and protein are called cadmium-binding proteins (CBPs), which have lower toxicity than inorganic cadmium (Lambert, 1983). In our body system, it is slowly excreted after combining with this proteins, such as albumin and metallothionein or negative ions of other molecules, especially the –SH group, without entering into metabolic process (Shim *et al.*, 2009). 13 and 13.5 KD CBP was isolated and purified from rice grain endosperm and phloem sap respectively. 4 QTLs for Brown rice Cd concentration (BRCdC) were detected, among which, 2 novel QTLs (qBRCdC9 and qBRCdC-12) present in chromosomes 9 and 12 in rice. Importantly, the QTL qBRCdC-12 is associated with the Cd translocation from shoot to brown rice grain (Guo *et al.*, 2019).

Rice grain may be contaminated with some fungal toxins like aflatoxin and many pesticides but the possibility of its occurrence is very less after post harvest processing.

Table 1: Probable strategies for bio-fortification of some nutritional compounds/parameters in rice.

Sl No.	Nutritional compounds/ parameters	Genetic approach	Agronomic	Inherent
1.	Phenolics	Over expression of PAL gene.	Biotic, abiotic and nutritional stress in rice field.	Pigmented rice eg. Rice genotypes Mamihunger, Chakhao etc
2	Flavonoids	Rice structural gene homologs encoding chalcone synthase (CHS), chalcone isomerase (CHI), flavanone 3-hydroxylase (F3H), flavonoid 3'- hydroxylase (F3'H), dihydroflavonol 4-reductase (DFR), and anthocyanidin synthase (ANS)	NA	Pigmented rice (red, black and purple rice)
3	Vitamin A	PSY (phytone synthase) and CRTI (phytoene desaturase) gene expression in rice.	NA	Golden rice
4	Vitamin E	Overexpression of tocopherol cyclase (VTE1) under the control of the 35S promoter in *A. thaliana* leaves was reported to result in a strong increase of tocochromanols. Other tocopherol synthesis genes are *HPPD (p*-hydroxyphenylpyruvate dioxygenase), *VTE2* (homogentisate phytyltransferase), and *VTE4 (*methyltransferase). The seven vitamin E biosynthesis geneOsHPPD, OsHPT ,OsHGGT, OsMPBQ MT1 OsMPBQ MT2, OsTC and OsγTMT are present in rice chromosome no.2,6,6,7,12,2,2 respectively (Chaudhary et al.,2009).	NA	NA
5	Oryzanol	NA	NA	Red and black rice
6	Selenium	Both phosphate transporter *OsPT2* and sulfate transporters *OsSultr1;2* may contribute to the uptake of selenate in rice.	Fertilization by sodium selenate (Na2SeO4) with 5 to 30 % availability. Plant uptake in the form of Selenate(SeO_4^{2-}) or selenite (SeO_3^{2-}) or in organic form as Se-amino acid, (Se-met). Higher dose of Se (> 80μM) is toxic to rice.	NA

7	Glycemic Index/load	A putative gene *sbe3-rs* (starch branching enzyme3-resistant starch) is associated with resistant starch content in grain. Three deleterious mutation/variants each in *GBSSI, SSIIa*, and *SSIIIa* have the potential to increase resistant starch content in rice.		
8	Vitamin B complex	**Vitamin B1** present in brown rice is about 2.9-6.1 μg/g and the two gene *THIC* (hydroxymethylpyrimidine phosphate synthase); *THI1* (thiazole biosynthetic protein) are responsible for thiamin biosynthesis in rice. **Vitamin B2 (Riboflavin)** is present in brown rice in the range of 0.4-1.4 μg/g. No report for genetic regulation of it's biosynthesis in rice. **Vitamin B3 (niacin):** is present in brown rice in the range of 35.0-53.0 μg/g. No report for genetic regulation of it's biosynthesis in rice. **Vitamin B5 (Pentothenic acid)** is present in brown rice in the range of 9.0-15.0 μg/g. Pantothenate synthetase gene was cloned from *Lotus japonicas* and *O. sativa.* **Vitamin B6** is present in brown rice in the range of 5.0-9.0 μg/g. Overexpression of Arabidopsis pyridoxal phosphate synthase 1 (PDX1) and pyridoxal phosphate synthase 2 (PDX2) genes enhances vitamin B6 level in *A. thaliana.* **Vitamin B7** (biotin) is present in brown rice in the range of 0.04-0.1 μg/g. Heterologous expression of *E. coli* diaminopimelate amino transferase or biotin synthase gene. **Vitamin B9 (folate)**The amounts of folate ranged from 0.1 to 0.5 mg/g in brown rice and reduced by 60%- 67% in milled rice.It was reported that overexpression of GTP cyclohydrolase I and amino deoxychorismate synthase genes enhance the rice folate level significantly (Ghosh et al., 2019). But there is no report about the presence of other vitamins like Vit-B12, C, D, K in rice.	**Folic acid physical fortification:** 5 mins. sonication time, overnight air drying then 3h soaking in 800 ppm folic acid (Tiozon et al.,2020) enhance folate content in rice grain.	NA

9	Aroma compounds (eg. 2-AP, (E,E)-2,4-decadienal, (E)-2-decenal, (E)-2-nonenal, decanal, nonanal and octanal)	**2AP:** Reduced expression of betaine aldehyde dehydrogenase 2 (*badh2*) present in chromosome 8 and*glyceraldehyde-3-phosphate dehydrogenase* (*GAPDH*) while overexpression of *triose phosphate isomerase* (*TPI)* and Δ^1-pyrolline-5-carboxylic acid synthetase (*P5CS*) gene enhance 2AP level in grain. The genes for other compounds are yet to be established.	Mn (MnSO4) enhance 2AP content by up regulating the 2AP synthesizing gene (Li et al., 2016)	Rice cultivars Pusa basmati, and landraces like Gobindabhog, Ketakijoha, Geetanjali, Kalajeera etc.

Genetic Analysis of Grain Antioxidants and Colour Formation in Case of Pigmented Rice

For improvement of the phenolics, flavonoids, and antioxidant capacity, we must understand the genetic bases of the related traits. Jin *et al.* (2009) found via linkage mapping that phenolic content, flavonoid content, and antioxidant capacity were individually controlled by three QTLs. Only one QTL on chromosome 2 was shared by phenolic content and flavonoid content. Shao *et al.* (2011) identified QTLs for these traits via association mapping using a diverse set of rice germplasm including red rice and black rice. Four, six and six QTLs were found associated with phenolic content, flavonoid content, and antioxidant capacity, respectively. Among them, four QTLs for phenolic content were also shared for other two traits. Ra (i.e. Prp-b for purple pericarp) and Rc (brown pericarp and seed coat) were main-effect loci for rice grain color and nutritional quality traits. Association mapping for the traits of the 361 white or non-pigmented rice accessions (i.e. excluding the red and black rice) revealed marker (RM346) is associated with phenolic content. Pigmented rice accumulates anthocyanins (black rice) and proanthocyanidin (red rice), which are beneficial to human health. Genetically, the pericarp color of red rice was controlled by two complementary genes, Rc (brown pericarp) on chromosome 7 and Rd (red pericarp) on chromosome 1. When present together, these loci produce red seed color. Rc in the absence of Rd produces brown seeds, whereas Rd alone has no phenotype (Sweeney *et al.* 2006; Furukawa *et al.* 2007). A natural mutation in rc has reverted brown pericarp to red pericarp and resulted in a new, dominant, wild-type allele, Rc-g . The color of dark purple pericarp was also controlled by two complementary genes, Pb and Pp, located on chromosome 4 and 1, respectively. Wang and Shu (2007) mapped Pb gene and suggested that this gene may be Ra gene. Markers for these genes may be useful for pigmented rice breeding, especially useful if new rice expects to accumulate both anthocyanins and proanthocyanidin.

Black rice landraces exist in at least three subspecies of rice, i.e., indica, tropical japonica, and temperate japonica. Generally, black rice undergoes genetic structural changes by the duplication and the insertion of genomic fragments in *Kala4* gene of tropical japonica. In rice, *R/B* homolog genes had been reported to be responsible in the regulation of anthocyanin biosynthesis (Hu *et al.*, 1996). The red colour pericarp is ubiquitous in the grain of wild rice (*Oryza rufipogon*) varieties, which share common ancestors with cultivated rice, and the defect in *Rc* gene of wild rice was spread broadly into the most domesticated rice varieties (Sweeney *et al.*, 2006). A 14-bp deletion within the *Rc* gene, which induced a premature stop codon, resulted in the change of pericarp color from red to white. To date, only two loci for the red colour trait had been reported; Rc, which encodes a bHLH transcription factor, and Rd, which encodes dihydroflavonol-4-reductase (DFR) (Sweeney *et al.*, 2006; Furukawa *et al.*, 2007). The black (or purple) grain color has not been observed in any accessions of *O. rufipogon*. Thus, the black rice trait is most likely newly acquired, incorporated either during or after rice domestication. Previously, the 'Hong XieNuo' (a black rice cultivar) alleles of three loci, namely, Kala1, Kala3, and Kala4, were introgressed into an elite temperate japonica cultivar 'Koshihikari' (Maeda *et al.*, 2014). The resulting 'Black rice NIL' showed complete conversion of the white pericarp into black pericarp. It has been speculated that the Kala1 and Kala3 genes encode a DFR and anR2R3-Myb transcriptional factor, respectively. However, Kala4 acts as a main contributor of the black trait in rice.Kala4 encodes a bHLH transcription factor, which is a rice homolog of the maize R/B gene.

Table 2: Comparison of different nutritional or anti-nutritional elements /compounds in pigmented and white rice.

Sl No.	Nutritional or antinutritional elements/compounds	Pigmented rice	White rice	References
1	Vitamin E (µg/g DW)	46.93-65.18	56.11	Min *et al.*,2014
	Anthocyanin (µg/g DW)	1394.90	NA	
	Total Phenolics (µg/g DW)	329.20	NA	
	Antioxidant activity (DPPH) (mg trolox eq./g DW)	2.50-10.80	0.10-0.20	
2	Total Phenolics content (µg-GAE/g DW) (whole grain)	421.00 – 437.00 (whole grain)	167.00 (whole grain)	Shao. et.al.,2014
3	Total Phenolics (mg GAE/g)	1.65-7.31(red) 8.41-12.44(black)	1.08-2.51	Shen . *et al.*, 2009
	Total Flavonoids(mg RE/g)	1.08-1.90(red) 1.87-2.86(black)	0.88-1.70	
	Antioxidant capacity (ABTS) (mM TE/100g DW)	0.29-2.96(red) 2.25-5.53(black)	0.01-0.41	
4.	Protein(%)	9.01-9.67	7.44-8.37	Kaur et.al.,2018
	Starch(%)	70.00-72.00	74.00-82.00	
	Antioxidant activity(%)	46.55-56.36	12.93-18.53	
	Total Phenolic content(%)	1.19-1.92	0.57-0.94	

5.	Fe(mg/kg)	10.73-18.64	11.05-20.62	Shao et. al.,2018
	Zn(mg/kg)	21.02-32.13	14.95-50.40	
	Cu(mg/Kg)	2.11-4.03	1.33-7.10	
	Mn(mg/Kg)	27.74-50.93	15.86-35.16	
	Mg(mg/Kg)	1039.25-1518.13	755.63-1277.13	

Minerals Content in Pigmented Rice

Micronutrients (Zn, Mn, Cu and Fe) can be found in every cell where they play important roles in maintaining normal metabolic functions, proper fluid balance, blood pressure regulation, nerve transmission, and immune system. Minerals acts as a cofactor of different enzymes of our body. K and Mg were the most abundant minerals in whole rice grains, accounting for about 60%and 30% of total minerals, respectively. Mn and Zn showed higher content in black rice than in most of non-pigmented and red rice samples (Table 4).

Table 3: Some inorganic ions that serve as cofactors for enzymes.

Sl. No.	Ions	Enzymes
1	Cu2+	Cytochrome oxidase
2	Fe2+,Fe3+	Cytochrome oxidase, catalase, Peroxidise
3	K+	Pyruvate kinase
4	Mg2+	Hexokinase, glucose 6-phosphatase, pyruvate kinase
5	Mn2+	Arginase, ribonucleotidereductase
6	Mo	Dinitrogenase
7	Zn2+	Carbonic anhydrase, alcohol dehydrogenase, carboxypeptidases A and B

Among all the minerals, phosphorus content (2062.1 to 2529.7 mg/kg) in pigmented brown rice (*Chak-hao Angangba*, brown rice; *Chak-hao Poireiton*, purple rice and *Chak-hao Amubi*, black rice), was the highest, followed by potassium (1546.8 to 1843.6 mg/kg), sulphur (743.5 to 976.1 mg/kg), magnesium (377.2 to 387.6 mg/kg), calcium (77.6 to 136.2 mg/kg), iron (47.2 to 88.8 mg/kg), and zinc (34.9 to 53.9 mg/kg), while copper content (27.5 to 33.4 mg/kg) was the lowest. However, polishing led to significant declines (0.75 to 2.50 times)in the mineral contents of the pigmented rice. (Reddy *et al.*,2017).

Selenium and Scope in Rice Bio-fortification

Bio-fortification of crops can be done with two approaches- genetic modification through biotechnological methods or conventional breeding and agronomic intervention. But now-a days physical fortification of foods can also be done from the external source. There are many nutritional compounds which can be bio-fortified but the bio-abailability for human beings should also be considered seriously for research. Selenium is uncommon mineral but essential for various human metabolism. Methylated Se-compounds can be important in cancer prevention and this element can be found in animal foods as well as in the plants of *Allium* and *Brassica* genus. The antioxidant enzyme glutathion-peroxidase activity is dependent on Se. It is found in unusual amino acid Se-methylselenocysteine and derivative of γ-glutamile. Based on research results sodium selenate (Na_2SeO_4) was selected for agronomic fortification of crops and applied in soil. But the accumulation of Se in different plant parts are different and it is a researchable issue in case of rice.

Conclusion

Biofortification of staple crops that have low nutrient content is an attractive strategy to avert nutritional deficiency. However, the main drawback of rice biofortification is the site of accumulation of essential nutrients and micronutrients. There is many evidence that after polishing of the grains, the content of minerals, vitamins, antioxidant compounds are decreased substantially. Therefore, there is a scope to biofortify rice grain targeting the endosperm. But in nature, an equilibrium is exist during grain filling stages, resulting a rare possibility to accumulate all the nutritional compounds in one variety through biofortification.

Nutrients and micronutrients can be incorporated in grain through biofortification and physical fortification. One major drawback of physical fortification is the limited stability of the additives, for example, folate added to rice becomes more soluble at higher temperatures and is lost when the rice is boiled.A second disadvantage is that additives can also affect the quality of food, even sensory quality.Multivitamin corn (registered as the protected variety Carolight1 in Spain) was developed by transforming an elite white-endosperm South African inbred line with four genes representing three different vitamin biosynthesis pathways, increasing the levels of β-carotene, other carotenoids, vitamin C and folate. Therefore, same approach may be applied for vitamins biofortification in rice. The micronutrient Se plays an important role in different metabolic process. The most known Se-proteins in human body include glutathion-peroxidase (GSH-Px), thioredoxin reductase which act as antioxidative enzyme and redox reaction enzyme respectively. Besides, Se-met and Se-cys are the amino acids which are the component of some structural protein. Therefore, Se biofortification in rice will be the new dimension in the field of nutritional research.

References

Ahn, S.C., Chang, J.Y., Lee, J.S., Yu, H.Y., Jung, A.R., Kim, J.Y., Choi, J.W., Hong, Y.S., Yu, S.D., Choi, K. 2016. Exposure factors of cadmium for residents in an abandoned metal mine area in Korea. Environ. Geochem. *Health* **39,** 1059–1070.

Akter, K.F., Owens, G., Davey, D.E., Naidu, R 2005. Arsenic speciation and toxicity in biological systems. *Rev Environ Contam Toxicol,* **184:**97-149.

Akihisa, T., Yasukawa, K., Yamaura, M., Ukiya, M., Kimura, Y., Shimizu, N., *et al.* 2000. Triterpene alcohol and sterol ferulates from rice bran and their antiinflammatory effects. *J Agric Food Chem.* **48:** 2313–2319

Banterng, P. and Joralee, A.2014. Evaluation of black glutinous rice genotypes for stability of gamma oryzanol and yield in tropical environments. Turk J Field Crops 20(2), 142-149

Bergman, C. J. & Xu, Z. 2003. Genotype and environment effects on tocopherol, tocotrienol, and c-oryzanol contents of Southern US rice. Cereal Chem. 80, 446–449.

Bunzel, M., Allerdings, E., Sinwell, V., Ralph, J., & Steinhart, H. 2002. Cell wall hydroxycinnamates in wild rice (*Zizania aquatica* L.) insoluble dietary fibre. Eur. Food Res. Technol. 214, 482–488.

Butsat, S. and Siriamornpun, S. 2010. Antioxidant capacities and phenolic compounds of the husk, bran and endosperm of Thai rice. *Food Chem.***119**(2) 606-613.

Brand-Miller, J,, Hayne, S,, Petocz, P,, Colagiuri, S.2003. Low-glycemic index diets in the management of diabetes: a meta-analysis of randomized controlled trials. Diabetes Care. 26:2261–67.

Blancquaert, D., Van Daele, J., Strobbe, S., Kiekens, F., Storozhenko, S., De Steur, H., Van DerStraeten, D. 2015. Improving folate (vitamin B9) stability in biofortified rice through metabolic engineering. *Nat. Biotechnol* **33:** 1076- 1078.

Chakuton, K., Puangpronpitag, D., and Nakornriab, M. 2012. Phytochemical content and antioxidant activity of colored and noncolored thai rice cultivars. *Asian. J. Plant Sci.,* **11:** 285-293.

Cho, J.Y., Lee, H. J., Kim, G. A., Kim, G.D., Lee, Y. S., Shin, S. C., Park, K.H., Moon, J.H. 2012. Quantitative analyses of individual g-Oryzanol (Steryl Ferulates) in conventional and organic brown rice (*Oryza sativa* L.). *J. Cereal Sci.* 55 (2012) 337-343

Chandel, G., Banerjee, S., See, S., Meena, R., Sharma, D. J., and Verulkar, S. B. 2010. Effects of different nitrogen fertilizer levels and native soil properties on rice grain Fe, Zn and protein contents. Rice Sci. 17, 213–227. doi: 10.1016/ S1672-6308(09)60020-2

Chaudhary, N., Khurana, P. 2009. Vitamin E biosynthesis genes in rice: molecular characterization, expression profiling and comparative phylogenetic analysis. Plant Sci. **177:** 479-491

Dexter, P.B., 1998. Rice Fortification for Developing Countries. OMNI/USAID (NO. 15).

Dong, W., Cheng, Z., Lei, C., Wang, X., Wang, J., Wang, J., Wan, J. 2014. Overexpression of folate biosynthesis genes in rice (*Oryza sativa* L.) and evaluation of their impact on seed folate content. *Plant Food Hum Nutr.* **69**(4), 379-385.

FAO, WHO, 1998. Carbohydrates in human nutrition: report of a joint FAO/WHO expert consultation. *FAO Food Nutr. Pap.* 66, 1–140

Foster-Powell, K, Holt, S.H.A., Brand-Miller, J.C. 2002. International table of glycemic index and glycemic load values. *Am J Clin Nutr.* **76**(1):5–56.

Guo, J., Li, K., Zhang, X., Huang, H., Huang, F., Zhang, L., Wang, Y., Li, T., Yu, H. 2019. Genetic properties of cadmium translocation from straw to brown rice in low-grain cadmium rice (*Oryza sativa* L.) *line. Ecotoxicol. Environ. Saf.* **182**: 109422

Ha, T. Y., Ko, S. N., Lee, S. M., Kim, H. R., Chung, S. H., Kim, S. R., *et al.* 2006. Changes in nutraceutical lipid components of rice at different degrees of milling. *Eur J Lipid Sci Tech,* **108**(3), 175–181.

Ismail, M., Al-Naqeeb, G., Mamat, W.A.A., Ahmad, Z.. 2010. Gamma-oryzanol rich fraction regulates the expression of antioxidant and oxidative stress related genes in stressed rat's liver. *Nutr Metab* **7**(23):1–13.

Iqbal, S., Bhanger, M. I., and Anwar, F. 2005. Antioxidant properties and components of some commercially available varieties of rice bran in Pakistan. *Food Chem.* **93:** 265–272

Jenkins, D,, Wolever, T., Taylor, R.H., *et al.* 1981. Glycemic index of foods: a physiological basis for carbohydrate exchange. *Am J Clin Nutr.*, **34**(3):362–6.

Juliano, B.O. 1985. Criteria and tests for rice grain qualities. In: Juliano BO (ed) Rice chemistry and technology. *American Association of Cereal Chemists Inc, St. Paul* 443–524

Juliano, B.O.1993. Rice in Human Nutrition, first ed. International Rice Research Inst.

Juliano, B.O., Bechtel, D.B. 1985. The rice grain and its gross composition. In: Juliano, B.O (ed.), Rice Chemistry and Technology, second ed. American Association of Cereal Chemistry, St. Paul, MN, pp. 17-57.

Jung, M.Y., Jeon, B.S., Bock, J.Y. 2002. Free, esterified, and insoluble-bound phenolic acids in white and red Korean ginsengs (Panax ginseng C.A. Meyer). *Food Chem.* **79:**105–11.

Jiang, G., Xu, L., Song, S., Zhu, C., Wu, Q., Zhang, L., Wu, L. 2008. Effects of long-term lowdose cadmium exposure on genomic DNA methylation in human embryo lung fibroblast cells. *Toxicology,* **244:** 49–55.

Kasim, A.B., Edwards Jr., H. 1998. The analysis for inositol phosphate forms in feed ingredients. *J. Food. Sci Agri.* **76:**1–9.

Kawakatsu, T. and Takaiwa, F. 2010. Cereal seed storage protein synthesis: fundamental processes for recombinant protein production in cereal grains. *Plant Biotechnol. J.* **8**: 939 -953 .

Ke, S., Cheng, X.Y., Zhang, N., Hu, H.G., Yan, Q., Hou, LL, *et al.* 2015. Cadmium contamination of rice from various polluted areas of China and its potential risks to human health. *Environ Monit Assess,* **187:** 408.

Khatoon, S. & Gopalakrishna, A. G. 2004. Fat-soluble nutraceuticals and fatty acid composition of selected Indian rice varieties. *J Am Oil Chem Soc*, **81:** 939–943.

Kumar, A., Sahoo, U., Baisakha, B., Okpani, O. A., Ngangkham, U., Parameswaran, C., Basak, N., Kumar, G., and Sharma, S. G. 2018. Resistant starch could be decisive in determining the glycemic index of rice cultivars. *J. Cereal Sci.* **79:**348-353.

Lambert, R.J.(1983) Tissue Residues and Toxicities of Inorganic and Protein-Bound Cadmium in Rats. University of Illinois at Urbana-Champaign.

Li, M., Ashraf, U., Tian, H., Mo, Z., Pan, S., Anjum, S. A., Duan, M., Tang, X.2016. Manganese-induced regulations in growth, yield formation, quality characters, rice aroma and enzyme involved in 2-acetyl-1-pyrroline biosynthesis in fragrant rice. *Plant Physiol Biochem.* **103:** 167-175

Li, K.T., Moulin, M., Mangel, N., Albersen, M., Verhoeven-Duif, N.M., Ma, Q. 2015. Increased bioavailable vitamin B6 in field-grown transgenic cassava for dietary sufficiency. *Nat Biotechnol.* **33:**1029-1032.

Liu, Z., Cheng, F., Zhang, G. 2005. Grain phytic acid content in japonica rice as affected by cultivar and environment and its relation to protein content. *Food Chem,* **89:** 49–52.

Luevano, J., Damodaran, C. 2014. A review of molecular events of cadmium-induced carcinogenesis. *J. Environ. Pathol. Toxicol. Oncol.* **33:** 183–194.

Meharg, A,A., Norton, G., Deacon, C., Williams, P., Adomako, E.E., Price, A. 2013. Variation in rice cadmium related to human exposure. *Environ Sci Technol,* **47**:5613 -8.

Mondal, D., Periche, R., Tineo, B., Bermejo, L.A., Rahman, M.M., Siddique, A.B., Rahman, M.A., Solis, J.L., Cruz, G.J.F. 2020. Arsenic in Peruvian rice cultivated in the major rice growing region of Tumbes river basin. *Chemosphere,* **241:** 125070.

Miller, J.B., Pang, E., Bramall, L. 1992. Rice: a high or low glycemic index food? *Am J Clin Nutr.*, **56**(6):1034–6.

Nasaretnam, K. 2005. Palm tocotrienols and cancer. In D. Bagchi & H. G. Preuss (Eds.), Phytopharmaceuticals in cancer chemoprevention, (pp. 481–490). London: CRC Press.

Nicolosi, R. J., Rogers, E. J., Ausmann, L. M., & Orthoefer, F. T. 1993. Rice bran oil and its health benefits. In W. E. Marshall & J. I. Wadsworth (Eds.), Rice science and technology (pp. 421–431). New York: Marcel Dekker.

Orthoefer, F. T. 2005. In: Bailey's Industrial Oil and Fat Products, Sixth Edition, Six Volume Set. Edited by Fereidoon Shahidi. Copyright, John Wiley & Sons, Inc, 465 -489.

Pascual, C. D. S. C. I., Massaretto, I. L., Kawassaki, F., Barros, R. M. C., Noldin, J. A., & Marquez, U. M. L. 2013. Effects of parboiling, storage and cooking on the levels of tocopherols, tocotrienols and γ-oryzanol in brown rice (*Oryza sativa* L.). *Food Res. Int.,* **50**(2), 676–681.

Phimolsiripol, Y, Mukprasirt, A., Schoenlechner, R. 2012. Quality improvement of rice-based gluten-free bread using different dietary fibre fractions of rice bran. *J. Cereal Sci.*, **56:** 389-395.

Pradhan S.K, Pandit E, Pawar S., Pradhan A., Behera L., Das S.R., Pathak H. 2020. Genetic regulation of homeostasis, uptake, bio-fortification and efficiency enhancement of iron in rice.*Environ. Exp. Bot.*. https://doi.org/10.1016/j.envexpbot.2020. 104066

Rafiq, M., Aziz, R., Yang, X., Xiao, W., Rafiq, M., Ali, B., *et al.* 2014. Cadmium phytoavailability to rice (*Oryza sativa* L.) grown in representative Chinese soils. A model to improve soil environmental quality guidelines for food safety. *Ecotoxicol. Environ. Saf.* **103:**101–107.

Rahman, M.M., Owens, G., Naidu, R. 2009. Arsenic levels in rice grain and assessment of daily dietary intake of arsenic from rice in arsenic-contaminated regions of Bangladesh-implications to groundwater irrigation. *Environ. Geochem. Health* **31**(1):179–187.

Roychowdhury, T. 2008a. Impact of sedimentary arsenic through irrigated groundwater on soil, plant, crops and human continuum from Bengal delta: special reference to raw and cooked rice. *Food Chem. Toxicol.* **46**(8):2856–2864

Salgado, J. M., Anderson, G. C, de Oliveira, Débora N. M., Carlos M. Donado-Pestana, Candido Ricardo Bastos and Fernanda Klein Marcondes 2010. The Role of Black Rice (*Oryza sativa* L.) in the Control of Hypercholesterolemia in Rats. *J. Med Food*, 1355-1362.

Song, J.Y. , Lee, M-C., Lee, D-J., Lee, G-A., Park, H-J., Lee, J-R., Choi, Y-M., Lee, S. K., Jung, Y., Cho Y-G. 2013. Analysis and comparison of the γ-oryzanol content based on phylogenetic groups in Korean landraces of rice (*Oryza sativa* L.) *Plant Breed Biotechnol.* **3**(1),58-69

Sun, R. C., Sun, X. F., & Zhang, S. H. 2001. Quantitative determination of hydroxycinnamic acids in wheat, rice, rye, and barley straws, maize stems, oil palm frond fiber, and fast-growing poplar wood. *J.Agric. Food Chem.* **49:** 5122– 5129.

Sun, R. C., Sun, X. F., Wang, S. Q., Zhu, W., & Wang, X. Y. (2002) Ester and ether linkages between hydroxycinnamic acids and lignins from wheat, rice, rye, and barley straws, maize stems, and fast-growing poplar wood. *Ind. Crops Prod.*, **15**: 179– 188.

Shim, J.A., Son, Y.A., Park, J.M., Kim, M.K.(2009) Effect of Chlorella intake on Cadmium metabolism in rats. *Nutr. Res. Pract.* **3**(1): 15–22.

Somaratne, G. M., Prasantha, B. D. R., Dunuwila, G. R., Chandrasekara, A.,Wijesinghe, D., & Gunasekara, D. C. S. 2017. Effect of polishing on glycemic index and antioxidant properties of red and white basmati rice. *Food Chem.*, **237:**716-723.

Tang, Y., Cai, W., Xu, B. 2015. From rice bag to table: Fate of phenolic chemical compositions and antioxidant activities in waxy and non-waxy black rice during home cooking. Food Chem. http://dx.doi.org/10.1016/j.foodchem.2015.02.001

Tiozon Jr., R.N., Camacho, D.H., Bonto, A.P., Oyong, G.G., Sreenivasulu, N. 2020. Efficient fortification of folic acid in rice through ultrasonic treatment and absorption, *Food Chem.* doi: https:// doi.org/10.1016/j.foodchem.2020.127629

Vichapong, J., Maliwan, S., Voranuch, S., Prasan, S., Supalax, S. 2010. High performance liquid chromatographic analysis of phenolic compounds and their antioxidant activities in rice varieties. *LWT-Food Sci Technol.* **43:**1325–1330.

Wilson, T.A., Nicolosi, R.J., Woolfrey, B., Kritchevsky, D. 2007. Rice bran oil and oryzanol reduce plasma lipid and lipoprotein cholesterol concentrations and aortic cholesterol ester accumulation to a greater extent than ferulic acid in hypocholesterolemic hamsters. *J Nutr Biochem,***18**:105–12.

Xie, P.P., Deng, J-w., Zhang, H-min., Ma, Y-hua., Cao, D-j , Ma, R-xiao, Liu, R-jing, Liu, C, Liang, Y-g (2015)Effects of cadmium on bioaccumulation and biochemical stress response in rice (*Oryza sativa* L.) *Ecotoxicol. Environ. Saf.* **122**: 392-398.

Yasukawa, K., Akihisa, T., Kimura, Y., Tamura, T. and Takido, M. 1998. Inhibitory e#ect of cycloartenol ferulate, a component of rice bran, on tumor promotion in two-stage carcinogenesis in mouse skin. *Biol. Pharm. Bull.*, **21:** 1072-1076.

Yu, J., Vasnathan, T., Temelli, F. 2001. Analysis of phenolic acids in barley by high performance liquid chromatography. *J.Agric Food Chem.*, **49:**4352–8.

Zhang, M.W., Zhang, R. F., Zhan, F. X. and Liu, R.H. 2010. Phenolic Profiles and Antioxidant Activity of Black Rice Bran of Different Commercially Available Varieties. *J. Agric Food Chem.* **58**(13),7580–7587.DOI: 10.1021/jf1007665.

Jin L, Xiao P, Lu Y, Shao Y, Shen Y, Bao J. 2009. Quantitative trait loci for brown rice color, phenolics, flavonoid contents, and antioxidant capacity in rice grain. *Cereal Chemistry.* **86**(6):609-15.

Shao Y., Jin L., Zhang G., Lu Y., Shen Y., Bao J. 2011. Association mapping of grain color, phenolic content, flavonoid content and antioxidant capacity in dehulled rice. *Theor Appl Genet.* **122:**1005–1016. DOI 10.1007/s00122-010-1505-4.

Sweeney MT, Thomson MJ, Pfeil BE, McCouch SR 2006. Caught red-handed: Rc encodes a basic helix–loop–helix protein conditioning red pericarp in rice. *Plant Cell* **18:**283–294.

Wang C & Shu Q 2007. Fine mapping and candidate gene analysis of purple pericarp gene *Pb* in rice (*Oryza sativa* L.)*Chinese Science Bulletin.* **52**:3097–3104.

Furukawa, T., Maekawa, M., Oki, T., Suda, I., Iida, S., Shimada, H., *et al.* 2006. The Rc and Rd genes are involved in proanthocyanidin synthesis in rice pericarp. *Plant J.* **49:** 91–102. doi: 10.1111/j.1365-313X.2006.02958.x

Hu, J., Anderson, B., and Wessler, S.R. 1996. Isolation and characterization of rice R genes: evidence for distinct evolutionary paths in rice and maize. *Genetics* **142:** 1021–1031.

Maeda, H., Yamaguchi, T., Omoteno, M., Takarada, T., Fujita, K., Murata, K., *et al.* 2014. Genetic dissection of black grain rice by the development of a near isogenic line. *Breed. Sci.* **64:**134–141. doi: 10.1270/jsbbs.64.134.

7

Transgenic and Genome Editing Approaches for Biofortification In Rice

Introduction

The transgenic approach can be an effective alternative for the development of biofortified crops when there is limited genetic variation for the targeted nutrient content among the germplasm (Zhu *et al.*, 2007). Transgenic approaches have been used for the simultaneous integration of genes involved in the enhancement of micronutrient concentration as well as reduction in the concentration of antinutrients, resulting in enhancing the bioavailability of nutrients in plants. Development of biofortified crops employing the transgenic approaches, involves a significant amount of time, effort, and resources during the development stage, but it is a cost-effective and sustainable approach in a long run (Garg *et al.,* 2018). For its wide consumption, rice has been targeted to address the global challenge of malnutrition. Vitamin deficiency can be considered as one of the significant challenges that affect poor section of populations due to low affordability. Development of 'Golden Rice' with provitamin A (beta-carotene) was recognized as an important breakthrough in this direction by expressing genes encoding *PSY* and carotene desaturase (Datta *et al*., 2003). Folic acid (Vitamin B9) is important during pregnancy and reducing anemia. Rice was genetically modified to increase folate content (up to 150-fold) by overexpressing genes encoding *Arabidopsis* GTP-cyclohydrolase I (GTPCHI) and aminodeoxy-chorismate synthase (ADCS) (Blancquaert *et al*., 2015). Rice has also been targeted to address the global challenge of iron deficiency anemia. Multiple reports indicated an enhance in iron content by overexpressing a number of genes (Lee *et al*., 2009). Similarly, zinc content was also raised in rice by overexpressing *OsIRT1* and mugineic acid biosynthesis genes from barley (*HvNAS1*, *HvNAS1*, *HvNAAT-A*, *HvNAAT-B*, *IDS3)* (Lee and An 2009; Masuda *et al*., 2008). Improvement in quality protein has been addressed by targeting essential amino acid content in rice by expressing seed-specific genes. An essential fatty acid, α-linolenic acid has been enhanced in rice by expressing the soybean omega-3 fatty acid desaturase (FAD3) gene (Anai *et al.*, 2203) that increases the amount of polyunsaturated fatty acid that can help in the reduction of bad cholesterol levels in the body and improve human nutrition. Flavonoids are associated with antioxidant activity and their content in rice has been enhanced by expressing maize C1 and R-S regulatory genes (Shin *et al.*, 2006) and phenylalanine ammonia-lyase and chalcone synthase (*CHS*) genes (Ogo *et al.*, 2013). To address the challenge of overnutrition and obesity, the content of less digestible and resistant amylose starch has been enhanced by the expression of antisense waxy genes (Leu *et al.*, 2003). Besides introducing micronutrients, the expression of functional human milk protein (lactoferrin) in rice grains has opened the possibility of creating value-added cereal-based ingredients that can be introduced into infant formula and baby food (Lee *et al.*, 2010). In this chapter, we are mainly focusing on the updates of transgenic research and development in important biofortification traits such as vitamins (A, B9), minerals (Fe, Zn), and protein quality (lysine) in rice.

Development of Golden Rice 1 and Golden Rice 2

To produce beta-carotene in rice endosperm, initially two scientists Ingo Potrykus worked at the Swiss Federal Institute of Technology in Zurich, Switzerland, and Peter Beyer worked at the University of Freiburg, in Freiburg, Germany started their work on rice. Generally, carotenoid production in plants needs the activity of four different enzymes. It started with GGPP (Geranylgeranyl Diphosphate). As cloning and transformation of multiple genes create difficulty, here, instead of *PDS* (phytoene desaturase) and *ZDS* (zeta-carotene desaturase), this group of scientists used a gene that can perform the function of both enzymes PSY and ZDS. Initially, they cloned three different genes, named *PSY* (Phytoene synthase), *CrtI* (Carotene desaturase), and *LCY* (Lycopene cyclase) to start the beta-carotene biosynthetic pathway in rice endosperm. The genes *PSY* and *LCY* were cloned from daffodil (*Narcissus pseudonarcissus*) another important gene was cloned from a soil-borne bacterium *Erwinia uredovora*, which codes for *CrtI* (Carotene desaturase) gene. For the successful transformation of these genes in plants, they cloned these genes in between the left and right borders of T-DNA. Finally, constructs were co-transformed into rice embryonic calli employing *Agrobacterium*-mediated transformation. The US Rockefeller Foundation supported their collaboration and the scientists and their collaborators first succeeded in expressing beta-carotene in rice in 1999, and they published the results in 2000 (Yo *et al.* 2000). Later, Datta *et al.* (2003), similarly transformed different members of *indica* (IR64, IR68144, BR29, Nang Hong Cho Dao, Mot Bui, Immyeobaw, and IR68899B) rice genotypes by particle bombardment method and developed different *indica* members of golden rice. Initially, it was believed that 3 genes are involved and essential for beta-carotene production in rice endosperm, but furthermore research works in this regard identified *LCY* gene is also expressed in wild rice endosperm and needs not to be over-expressed in rice endosperm. Therefore, only two genes are essential for the production of beta-carotene in rice endosperm *PSY* and *CRTI.* Initially, lines are developed by over-expressing *PSY, CRTI,* and *LCY* genes under the regulation of endosperm-specific promoter Gt1 (glutelin promoter), which produced 1.6 µg/g carotenoids in rice endosperm. The golden rice developed by over-expressing 3-genes are generally known as golden rice 1 or the first generation of golden rice and it is not sufficient to fulfill the overall demand for vitamin-A in the human body. Therefore, it was not very helpful to combat vitamin-A deficiency. To further improve the overall carotenoid content of golden rice, Paine *et al.* (2005) started work and produce the next generation of golden rice or golden rice 2. They identified that the gene *PSY*, which was originally isolated from *Narcissus pseudonarcissus*, could be the limiting factor for the accumulation of carotenoids in rice endosperm. After screening different *PSY* genes from different organisms, Paine *et al.* (2005) identified an efficient PSY gene from *Zea mays*, which can accumulate the highest amount of beta-carotene in rice endosperm. Therefore, it provided an effective solution to combat the previous problem and over-expressing lines containing *ZmPSY* gene were able to accumulate 37 µg/g carotenoid, which was almost 23 times higher than the previous golden rice.

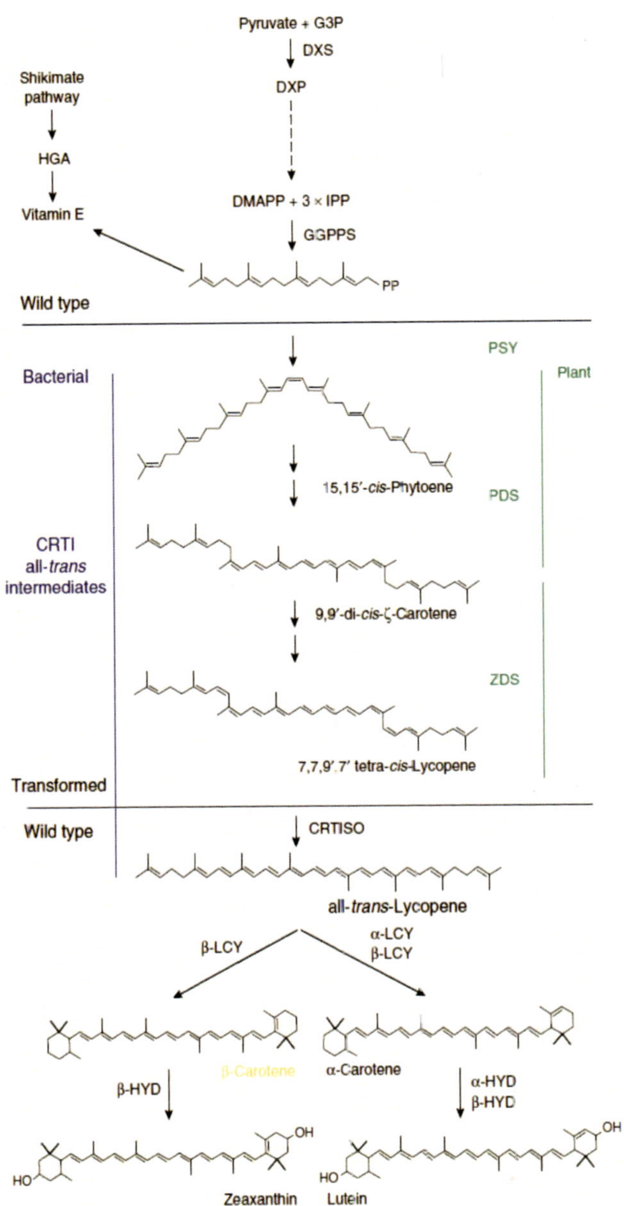

Fig. 1: β-Carotene biosynthesis pathway related to golden rice. (*Note*: GGPP, geranylgeranyl-diphosphate; DXS, 1-deoxy-Dxylulose-5-phosphate synthase; IPP, isopentenyl-diphosphate; DMAPP, dimethylallyl-diphosphate; GGPPS, GGPP synthase; HGA, homogentisic acid; PSY, phytoene synthase; PDS, phytoene-desaturase; ZDS, z-carotene-desaturase; CRTISO, carotene cis-trans-isomerase; LYC, lycopene cyclases; HYD, hydroxylases). Reproduced from Al-Babili and Beyer, 2005, Trends in Plant Science, https://doi.org/10.1016/j.tplants.2005.10.006.

Current Status of Golden Rice

To secure basic immunity for human health and fight against vitamin-A deficiency, initially golden rice project was started. The principal reason behind this project is to serve vitamin-A to the malnourished people who subsist on a bowl of rice. Keeping this in mind, initial work was started in the year 1999 and subsequently the first generation of golden rice carrying the successfully expressed carotenoid biosynthetic genes in rice endosperm was developed which could accumulate carotenoids in the endosperm (Yo *et al.*, 2000). Finally, by their intensive effort, the dream of development of the first generation golden rice became true. Later, Datta *et al.* (2003) developed some transgenic lines in *indica* background that synthesize beta-carotene in the endosperm. Though they have successfully expressed and started carotenoid biosynthesis in rice endosperm, however, the amount present in the endosperm is low (nearly1.6 µg/g). Therefore, it was not very helpful in fulfilling the demand for carotenoids in the human body. Later, this issue was sorted by Paine *et al.* (2005), in their work, they investigated the functionality of different *PSY* genes from different organisms and identified *PSY* genes from maize that can provide a solution for this purpose. By transforming two rice genotypes they developed the second generation of golden rice or golden rice 2. Based on their work, it was found that the expression of maize *PSY* in combination with *CrtI* can accumulate almost 37 µg/g carotenoid in rice endosperm. At this point, the first generation of golden rice shifted to the second generation, and scientists accepted this next generation of golden rice or golden rice 2, which was more helpful to meet the overall vitamin-A demand in the human body.

Although the development of golden rice 2 was completed in 2005, till now, the main purpose for developing golden rice 2 is not fulfilled. This was developed by a horizontal gene transfer process and by applying different techniques of transformation. It is designated as a genetically modified plant. Initially, it was grown only in greenhouses with critical monitoring. Till now, this project is not getting permission for its field production in India and is generally produced in greenhouses or a few areas in the field with critical monitoring. At present, golden rice is still in the developing phase and must undergo bio-safety measures before its introduction in the market (Potrykus, 2001; Mayer and Potrykus, 2011; Zeigler, 2014). Once it is approved by the food safety regulatory bodies will be distributed for field production.

The prime reason behind this, the natural genetic backbone of rice was compromised by the introgression of foreign genes; therefore, different authorities and governments did not encourage this transgenic development process wilfully and further creating difficulties for its field release. Though it has some negative impacts, it has to be understood clearly, beta-carotene-rich rice endosperm is not possible by conventional breeding. In case of conventional breeding, a better genotype was identified at first by the screening of different wild or landrace accession of rice or other plants, then it was crossed with popular/ high yielding variety to develop a new variety that contains the desirable traits. In case of rice, as the beta-carotene biosynthesis process was blocked in the endosperm, screening followed by identification of a better genotype is not a solution here. Therefore, the transgenic approach or modern molecular tools are the only way to fulfil the demand for vitamin-A-rich rice. To fulfil the nutritional requirement of poor people and to grow golden rice in the field, a thorough investigation of the nutritional value of the seed, overall production of rice, and detailed metabolic and proteomic activity of compounds, enzymes, and proteins has to be analyzed properly (Gayen *et al.*, 2016). As human health is involved here, negligence in any safety measures can be severe and non-recoverable.

Based on the reports, *javanica*, *japonica,* and *indica* rice cultivars have been used to develop golden rice, though their biochemical and physiological aspects need to be understood further. Therefore, the possible outcome of genetically modified crops combating nutritional deficiency and poverty is still a subject of controversy. In general, it can be said that golden rice 2 would reduce the nutritional crisis of people who are below the poverty line and will help reduce health-related issues. Its clinical and field trials before its marketization and production were started in different parts of the world. IRRI has also performed field trials in the year 2013-14 for the identification of better lines of beta-carotene enriched rice grain. Based on the most recent reports, BRRI (Bangladesh Rice Research Institute) developed BRRI Dhan 29 from BR29, which accumulates carotenoid in rice endosperm and this institution is very close to releasing these genotypes in the field for the production of carotenoid-enriched rice grains.

Enhancement of Fe through Transgenic Approach

Fe is mainly present in the aleuronic layer of the endosperm. After milling majority of the grain Fe is generally removed even in a genotype having high Fe content in brown rice. Therefore, through traditional breeding approaches the improvement of high-yielding varieties for Fe is not yet a feasible option. On the other hand, through the transgenic approach, Fe can be enhanced in the inner part of the endosperm by manipulation of Fe storage, uptake, translocation, influx, and enhancing bioavailability.

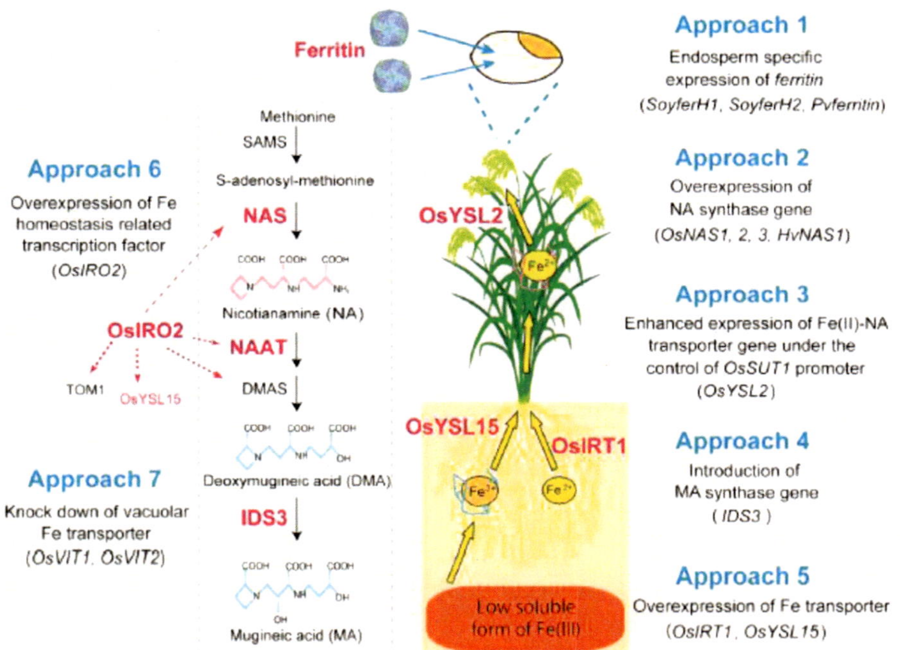

Fig. 2: Seven transgenic approaches to Fe biofortification of rice. The pathway inside the gray dashed-line rectangle shows the biosynthetic pathway for mugineic acid family phytosiderophores (MAs) in graminaceous plants. SAMS, S-Adenosyl-methionine synthase; NAS, NA synthase; NAAT, NA aminotransferase; DMA, 2'-deoxymugineic acid; DMAS, DMA synthase; IDS3, MA synthase (dioxygenase that catalyzes the hydroxylation of DMA and epiHDMA at the 2' position); Ferritin, iron storage protein; OsYSL2, Fe(II)-NA and Mn(II)-NA transporter; OsIRO2, Fe deficiency-inducible bHLH transcription factor related to Fe homeostasis in rice; OsIRT1, ferric transporter; OsYSL 15, Fe(III)-DMA transporter; TOM1, MA transporter. Rice lacks the two dioxygenase genes (IDS2 and IDS3) and secretes only DMA. **Approach 1:** Enhancing Fe accumulation in seeds by introducing the Fe storage protein, ferritin gene, SoyferH1, SoyferH2 or Pvferritin, under the control of endosperm-specific promoters. **Approach 2:** Enhancing Fe transport within the plant body by the overexpression of NAS. **Approach 3:** Enhancing Fe influx to seeds by expression of the Fe(II)-NA transporter gene OsYSL2 under the control of the OsSUTI promoter. **Approach 4:** Enhancing Fe uptake and translocation by introduction of the phytosiderophore synthase gene IDS3. **Approach 5:** Enhanced Fe uptake from soil by overexpression of the Fe transporter gene OsIRTI or OsYSL 15. **Approach 6:** Enhanced Fe uptake and translocation by overexpression of the OsIRO2 gene. **Approach 7:** Enhanced Fe translocation from flag leaves to seeds by knockdown of the vacuolar Fe transporter gene OsVITI or OsVITZ. The ferritin image was kindly provided by Dr. David S. Goodsell (Scripps Research Institute, La Jolla, CA, USA) and the RCSB PDB. (Partially reproduced from Masuda et al. 2013, Rice, 6: 40)

(i) ***Enhancement of Iron Storage in Rice***

Large and having 24 subunits, ferritin is an iron storage protein that can store up to 4500 iron atoms in a benign complex form. It also possesses ferroxidase activity (Theil 2003). Numerous plants have had their ferritin genes extracted and sequenced; however, the soybean ferritin gene, which is encoded by *SoyferH1* and *SoyferH2*, has received the greatest research attention (Kok *et al.* 2018). The soybean ferritin gene was thought to be a potential gene for iron biofortification in rice as the human gut can effectively absorb iron from the soybean ferritin iron complex (Theil 2011). Ferritin gene expression in rice was induced in numerous trials using endosperm-specific promoters, namely the rice glutelin (*OsGluB1*) and rice globulin (*OsGlb*) promoters. This resulted in a 3.7-fold increase in ferritin expression (Qu *et al.,* 2005; Paul *et al.*, 2015). Using molecular breeding techniques, high iron rice varieties, like IR 68144, were created by overexpressing the soybean ferritin gene, which has been shown to boost iron content in polished rice grain by 3.7 times (Vasconcelos *et al.*, 2003). Additionally, this characteristic is effectively transferred to the Indian m1ega rice variety Swarna by Paul *et al.* (2015), leading to 2.54 times more iron in milled rice grain than in control Swarna.

(ii) ***Enhancement of iron uptake***

As discussed previously, different iron phytosiderophore transporters play important roles in iron uptake in rice can be achieved by both reduction and chelation-based strategies (Kok *et al.*, 2018). In rice plants, nicotinamine synthase (NAS) and nicotinamine transferase are two key enzymes involved in the release of phytosiderophores with the help of TOM1 (transporter of MAs) transporter (Morrissey *et al.*, 2009). To achieve iron-fortified rice, overexpression of genes involved in MA biosynthesis was considered by many researchers. Overexpressing NAS was also attempted, as it catalyzes the synthesis of nicotianamine (NA) from S-adenosylmethionine (Wirth *et al.*, 2009). Three NAS genes have been identified from the rice genome- *OsNAS1*, *OsNAS2*, and *OsNAS3*, and overexpression of these genes gave satisfactory results and rice iron concentration was increased (Theil 2003). Overexpression of rice *OsNAS1*, *OsNAS2*, and *OsNAS3* was done by Johnson *et al.* (2011), while *OsNAS2* was overexpressed by Lee *et al.* (2012). Even the gene *HvNAS1* from barley was attempted to overexpress in rice by Masuda *et al.* (2009) to increase the iron concentration by double or more in polished rice.

(iii) ***Enhancement of iron translocation***

An array of MA genes is introduced in rice which improved the iron uptake process and its translocation (Majumder *et al.* 2019). These MA genes can activate *Iron Deficiency Specific* genes (*IDS2* and *IDS3*) in an iron-deficient environment which helps in enhancing iron uptake and translocation and combat iron deficiency in rice (Masuda *et al.*, 2013). The barley IDS genes can synthesize special types of MAs such as 3-epihydroxymugineic acid (epi-HMA), and 3-epihydroxy-20-deoxymugineic acid (epi-HDMA) from 2-deoxymugineic acid (DMA), the mechanism of which is lacking in rice per se (Kobayashi *et al.*, 2001). Moreover, the barley Fe^{3+}-MA complex has better stability in an acidic environment as compared to the rice Fe^{3+}-MA complex, which ultimately helps in more iron uptake and translocation (Von-Wiren *et al.* 2000). Hence, the introduction of barley IDS genes into rice through transgenic approaches could increase the iron uptake and its translocation in rice tissues (Masuda *et al.* (2008). It led to a ~1.5-fold rise in grain iron content in both brown and polished rice as compared to wild-type control plants.

In the case of plants, the transport of metals like Fe from different plant parts to grains is governed by different metal transporter genes which are present in the rice genome. In rice, vacuolar iron transporter genes (*OsVIT1* and *OsVIT2*) are one such class of transporters present in tonoplast which are involved in the transport of Fe^{2+} and Zn^{2+} ions into the vacuole (Kim *et al.*, 2006). Although expressed in all tissues of rice plants, it was found particularly active in flag leaves of rice and it was thought to be negatively regulating the final Fe grain accumulation in rice. Knocking down of OsVIT genes in rice resulted in a 1.4-fold increase in grain iron content and a significant decrease in Fe content in the flag leaf (Bashir *et al.*, 2013).

Further Kobayashi *et al.* (2013) identified two ubiquitin ligases *viz. OsHRZ1* and *OsHRZ2*, which are interesting new gene (*RING*)- and zinc-finger proteins and can bind to Fe^{2+} and Zn^{2+} ions and also possess ubiquitination activity. RNAi-mediated silencing of these genes resulted in 3.8-fold and 2.9-fold more Fe in brown and polished rice, respectively. *OsIRO2* is another basic helix-loop-helix (bHLH) transcription factor, which positively regulates grain iron accumulation in rice. Overexpression of this gene in transgenic rice resulted in ~2.0-fold greater iron build-up in brown rice as compared to the non-transgenic control lines (Ogo *et al.*, 2011).

(iv) *Enhancement of grain iron influx*

In the endosperm of rice, a total of eighteen distinct yellow stripe-like (YSL) genes are significant metal (iron)-chelator transporters (Koike *et al.*, 2004; Ishimaru *et al.*, 2010). The OsYSL2 iron nicotianamine transporter has been shown to particularly regulate iron inflow into rice endosperm. This group of transporters is engaged in the long-distance transport of iron-NA complex across the phloem (Koike *et al.* 2004; Schroeder *et al.*, 2013). Ishimaru *et al.* shown the significance of the OsYSL2 gene in the rice plant (2010). In comparison to control plants, disruption of the OsYSL2 gene reduced iron concentration by 18% in brown rice and 39% in polished rice (Ishimaru *et al.*, 2010). In a different experiment, the overexpression of the OsYSL2 gene in rice under the promoter of the sucrose transporter (OsSUT1) enhanced the concentration of iron in polished grain by almost four times (Ishimaru *et al.*, 2010). The overexpression of the OsYSL2 gene, which is driven by the OsSUT1 promoter, has been demonstrated to be an effective method for iron biofortification. In the future, a different promoter combination with more OsYSL genes might work better.

(v) *Increasing bioavailability of iron by modulating phytate level*

Phytic acid or myo-inositol hexaphosphahe is a compound that can chelate divalent cations like Fe^{2+}, Mg^{2+}, Zn^{2+}, or Ca^{2+} and thereby reduces the bioavailability of these ions in the human diet. Except for maize, in most of the cereals including rice, it is predominantly accumulated in the aleurone layer of the grain. The phytate content in the rice grain is one of the important determinants of the bioavailability of important minerals like iron and zinc. Over the years studies showed that some of the rice varieties/ germplasm accessions are having low phytate content (Kumar *et al.*, 2017). Also, there were a few attempts at rice by generating mutant varieties exhibiting a low phytic acid (lpa) phenotype in rice (Larson *et al.*, 2000; Kim *et al.*, 2008), which were effective to produce low grain phytate content but were having compromised yield and other agronomic traits. So, as an alternative strategy, transgenic lines were developed by manipulating the genes of key enzymes of the phytic acid biosynthetic pathway using RNAi technology (Ali *et al.* 2010; Karmakar *et al.*, 2019).

Ali *et al.* (2010) used RNA interference technology to silence the IPK1 gene using an Ole18 seed-specific promoter in order to create the Pusa Sugandhi II (PSII) indica rice cultivar. This allowed them to control the expression of the final stage critical enzyme, inositol-1,3,4,5,6-pentakisphosphate 2-kinase (IPK1) of phytic acid biosynthesis. The transgenic seeds showed a 3.85-fold down-regulation in IPK1 transcripts, which coincided leading to a notable decrease in phytate levels and a rise in inorganic phosphate (Pi) and without impairing the process of growth and development, collected 1.8 times more iron in the endosperm of transgenic rice plants. According to Karmakar *et al.* (2019), seed-specific phytic acid downregulation was 46.2%. RNA interference (RNAi)-mediated silencing of an ITPK homolog (OsITP/6K-1) in the *indica* rice variety, Khitish and discovered a 1.3-fold increase in iron in the seed with 1.6-fold zinc and 3.2-fold bioavailability of inorganic phosphate.

Transgenic Approaches for Increasing to Zn Content

Zinc is an important element and is used as a co-factor for over 300 enzymes and more than 1000 transcription factors (Palmgren *et al.*, 2008). There is limited natural diversity of grain Zn content in the cereals (Sharma *et al.*, 2015). For that reason, it is important to enhance the Zn concentration of cereal grains for attaining better metabolism and human nutrition. On the other hand, management of the Zn content in the cereal grain may be less straightforward than iron (Brinch-Pedersen *et al.*, 2007). Interestingly, Ozturk *et al.* (2006); reported a strong correlation between Fe, Zn, and grain protein content in rice. Gpc-B1 is a wheat quantitative trait locus that is related to improved grain protein content and also betters the Zn and Fe content (Uauy *et al.*, 2006). It was observed that an increase of 10–34% of Zinc, iron, Mn, and protein content in cultivated wheat by introducing Gpc-B1 locus which is isolated from the wild tetraploid wheat *Triticum turgidum* ssp. dicoccoides into different recombinant chromosome substitution lines, which is signifying the role of Gpc-B1 in the remobilization of protein, Zn, Fe, and Mn from the leaves to the grains (Distelfeld *et al.*, 2007). It is an important way to improve the Zn content in grain by up-regulation of genes that are involved in the translocation of Zn and mobilization with enhanced bioavailability of Zn without yield penalties. Many transporters are linked with cations that have been recognized in rice, but some of them have been categorized for the specificity of the substrate, expression pattern, and localization in the cell. Among the families of identified cation transporter, members of the ZIP (ZRT, IRT-related protein) and CDF (Cation diffusion facilitator) families are predominant as they are playing the most important role in the uptake of Zn and translocation. ZIP family protein *i.e* IRT1 protein contributes significantly to Zn uptake in root cells of *A. thaliana* (Korshunova *et al.*, 1999). An increase of 2–3-fold Zn content in paddy was observed by over-expressing the NA synthase (NAS) by introducing 35S enhancer elements (Lee *et al.*, 2009). Likewise, transgenic rice accumulation of 2–3-fold higher Zn content in polished rice grains was observed by expressing barley nicotianamine synthase gene *HvNAS1* under the control of the rice actin1 promoter (Masuda *et al.*, 2009). At IRRI, there are several thousand transformants of IR64 and IR69428 are identified with soybean or rice ferritin and rice nicotianamine synthase (NAS2) at IRRI by up-regulating the genetic constituents, and Zn and Fe content in those lines have surpassed the target level from the field trials. Therefore, the over-expression of NAS genes makes nicotianamine an interesting target for Zn biofortification. Furthermore, cereals biofortification with NAS alone or in combination with ferritin has a large probability to combat global human nutrient deficiency (Zheng *et al*., 2010; Lee *et al*., 2009). To know the pathway of iron and zinc in grain, satisfactory work is found mainly in many crop species like rice, wheat, barley, and maize. Despite numerous challenges like the root–shoot barrier and grain filling (Palmgren *et al*., 2008) wheat researchers build-up methods and resources developed in rice to make better enhancement in Zn content in wheat grain to found superior wheat varieties (Borrill *et al*., 2014).

In rice, iron and zinc uptake are regulated by members of the ZIP family transporter protein. OsZIP1, OsZIP2, OsZIP3, and OsZIP4 are linked with the uptake of metal and zinc homeostasis (Ishimaru *et al*., 2005, 2007), and OsZIP8 and OsZIP7a might encode zinc and iron transporter respectively (Yang *et al*., 2007). The OsZIP3 was overexpressed under both zinc-available and -deficient conditions, whereas *OsZIP1* gene was overexpressed under zinc-deficient conditions in rice (Ramesh *et al.,* 2003). Upregulation of *MxIRT* and *OsIRT* genes in GM rice was found in elevated iron and zinc concentration in rice grains (Lee *et al*. 2009; Tan *et al*., 2015), Expression of *AtIRT1* with Pvferritin and AtNAS1 genes for iron biofortification, and it was found that a 4.7-fold increase of iron and zinc concentration (Boonyaves *et al*., 2017). The transformation of a combination of four genes AtNAS1, AtIRT1, Afphytase, and Pvferritin,) in rice, resulted in iron and zinc build-up in GM rice increased in polished grain (Boonyaves *et al*., 2016). Multiple experiments have found that the upregulation of OsNAS genes enhances the concentrations of zinc and iron by several folds in rice grain (Johnson *et al*. 2011; Lee *et al*. ,2012). Upregulation of OsNAS1, OsNAS2, and OsNAS3 done by Johnson *et al*. 2011 found a two-fold increase in zinc content in rice seeds. Ali *et al*. (2013a) stated that RNAi-mediated silencing of MIPS gene of the phytic acid metabolism pathway enhanced zinc, magnesium, and calcium, concentration in milled rice grain along with iron. Ali *et al*. (2013b) stated that a similar result was found when another phytic acid metabolism gene, IPK1, was silenced and ITPK (OsITP/6K-1) was silenced by Karmakar *et al*., 2019). Paul *et al*., 2012 reported that the ferritin gene (Osfer2) was upregulated, and accumulated 1.37-fold and 2.09-fold of zinc and iron, respectively in PSII rice.

Transgenic Approach to Enhance Grain Protein and Essential Amino Acids

It is considered that the first limiting essential amino acid in rice is lysine, or Lys. Despite significant efforts to increase rice's Lys content by traditional breeding and genetic engineering, no products that meet market standards have been produced thus far. It is challenging to increase the levels of lysine in crops through traditional genetics and breeding methods because: (i) lysine synthesis is highly negatively regulated by a feedback inhibition loop that slows down the synthesis of the first enzyme in the lysine biosynthesis pathway, dihydro dipicolinate synthase (DHPS) (Box 1); and (ii) lysine is efficiently catabolised into the tricarboxylic (TCA) cycle, a pathway that is started by the bi-functional enzyme LKR/SDH (Box 1), which demonstrates both lysine-ketoglutarate reductase (LKR) and) and saccharopine dehydrogenase (SDH) activities (Karchi *et al.*, 1995).

Therefore, to increase essential amino acids modification of biosynthetic and catabolic fluxes has been employed. In rice and barley, the expression of the bacterial *DHPS* only slightly increased the content of free lysine (Lee *et al.* 2001). In rice, free lysine levels could be increased up to ~12-fold in leaves and ~60-fold in seeds by over-expression of *AK* and *DHPS* and silencing *LKR/SDH* by RNAi (Long *et al.* 2013). Long *et al.* (2013) showed that the level of LKR/SDH was significantly enhanced in the developing seeds of rice Asp kinase (AK)/DHPS overexpression lines. This means that the effect of transgene AK/DHPS was counterbalanced by the activity of lysine catabolism in sustaining a steady level of lysine. Overexpression of AK and DHPS only increased the free lysine level by 1.1-fold compared with the wild type; however, the LKR-RNAi line showed a 10-fold increase, while combined expression of AK and DHPS and interference of LKR/SDH to achieve both metabolic effects meant a substantial increase in the free lysine content to 60-fold. Nevertheless, it is by careful engineering of both of the above enzymes (DHPS and LKR/SDH) that Yang *et al.* (2016) have now developed two pyramid transgenic lines with free lysine content elevated to 25-fold compared to wild type without changing the plant phenotype. Taking this one step further, Yang *et al.* (2016) show a perfect example of meaningful enhancement of lysine levels in transgenic rice by using a combined enhancement of lysine synthesis and suppression of its catabolism.

Box 1. Biosynthesis and catabolism of lysine in plants

The pathways leading to and from lysine (biosynthesis/catabolism) also highlight targets for engineering. Enzymes are indicated in purple text. The dashed arrow indicates several non-specified enzymatic reactions. Abbreviations: AK, Asp kinase; DHPS, dihydrodipicolinate synthase; LKR, lysine ketoglutaric acid reductase; SDH, saccharopine dehydropine dehydrogenase; ASD, aminoadipic-semialdehyde dehydrogenase.

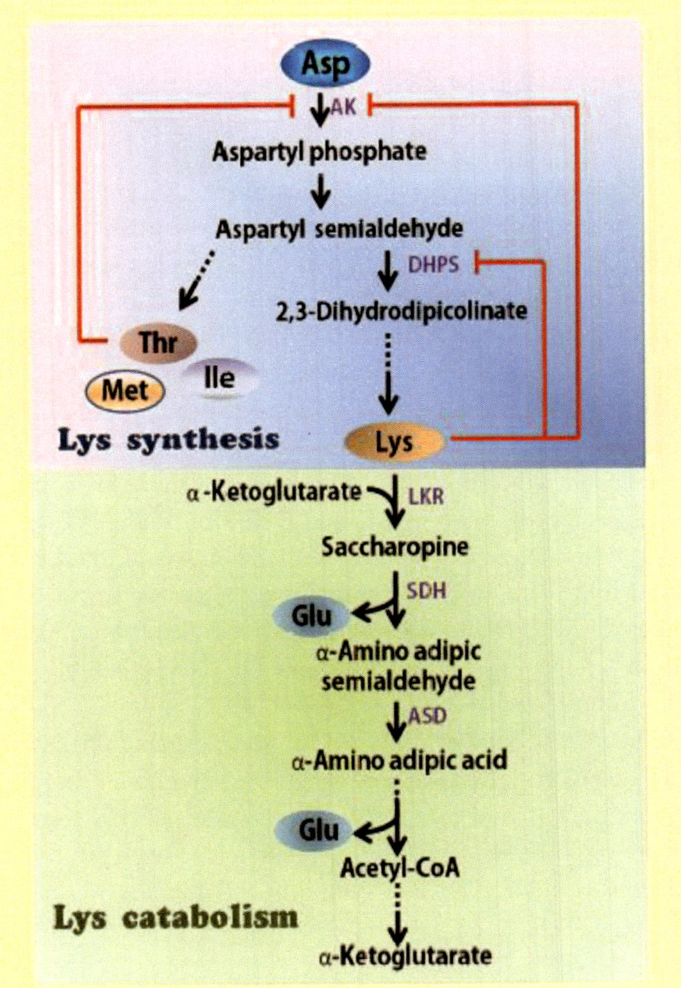

Fig. 3: Biosynthesis and catabolism of Lysine in Plants. The figure is reproduced from Galili, 2016, Journal of Experimental Botany, doi: 10.1093/jxb/erw254.

Wang W, Galili G. 2016. Transgenic high-lysine rice – a realistic solution to malnutrition? Journal of Experimental Botany, Vol. 67 No. 14 pp. 4009–4011, 2016, doi:10.1093/jxb/erw254

Making transgenic plants by over-expressing the genes that code for proteins with increased ratios of critical amino acids is another tactic. To increase lysine content, at least three lysine-rich genes have been identified. Endogenous rice genes RLRH1 and RLRH2 were identified; these genes encode proteins with lysine content of 14.7% and 20.6%, respectively, in their amino acid composition (Wong *et al.*, 2015). Transgenic plants demonstrated a 35% increase in lysine content when they were utilised to boost rice's lysine content (Wong *et al.*, 2015). Using an endosperm-specific GLUTELIN1 promoter (GT1), Liu *et al.* (2016) produced a LYSINE-RICH PROTEIN gene (LRP) from *Psopho carpustetragonolobus* (L.) DC in Peiai 64S (PA64S), an elite photoperiod-thermosensitive male sterility (PTSMS) line.

Western blot examination verified the foreign LRP protein's expression. Comparing the transgenic rice seeds to the wild-type, the Lys level rose by almost 30%, and the total amount of other amino acids increased as well. Three generations of persistent amino acid analysis revealed that transgenic rice seeds have a noticeably higher Lys content. In addition, the hybrid of the transgenic plants had a about 20% higher Lys content than the hybrid control. They tracked the transcript abundance of numerous genes encoding important enzymes involved in amino acid metabolism during the grain-filling stage. The findings revealed that decreased amino acid catabolism was the cause of the buildup of amino acids in transgenic plants.

When compared to the wild type, the genetically modified rice had undesirable grain morphologies; however, the hybrid derived from it demonstrated no or little detrimental impacts on grain. The transgenic plant seeds exhibited a considerable increase in Lys content due to the endosperm-specific expression of foreign LRP. This rise was found to be stable throughout three generations of testing. Lys content in the seeds of the hybrid of the transgenic plants also increased significantly. These findings suggested that increasing Lys levels in rice could benefit from the development of LRP in rice seeds.

Not only does rice lack lysine, but it also lacks threonine. Therefore, it is crucial to raise the lysine and threonine content of rice grains in order to further improve the quality of protein. Jiang *et al.* (2016) combined native rice genes with lysine (K)/threonine (T) motif (TKTKK) coding sequences to artificially synthesise two additional genes. They were assigned as *TKTKK1* and *TKTKK2*, and lysine/threonine makes up 73.1% and 83.5% of the encoded proteins, respectively. To create transgenic plants, these two genes—which were independently inserted into the rice genome and controlled by the 35S promoter—were introduced. When compared to the wild type control, data showed that overexpressing *TKTKK1* produced stable proteins with expected molecular weight and that transgenic rice seeds significantly increased the content of total amino acids, crude protein, lysine, and threonine by 33.87%, 21.21%, 19.43%, and 20.45%, respectively. Transgenic rice seeds overexpressing *TKTKK2* also showed significant improvement. However, transgenic seeds expressing a tandem array of these two novel genes showed a modest improvement in both the quantity and quality of protein. These findings offer a foundation and a different approach for using synthetic biology to further increase the quantity and quality of protein in other crops or vegetable plants.

Apart from conventional breeding transgenic approach also followed to develop lysine rich rice. A *japonica* paddy rice variety, "Heijing 5," was found suitable in Uppsala, Sweden, after several years' adaptation under a simple plastic cover when the temperature was below 10°C. Uppsala-adapted "Heijing 5" had high protein content (12.6%) in brown rice grain and was recommended for children complied with all dietary requirements determined by the Europeon Union and other countries for better nutrition. Heijing 5 generally produced grain yield of around 5100 kg/ha (Fei *et al.*, 2020).

Transgenic Rice for Vitamins

Overexpression of folate biosynthetic genes was found quite effective in enhancing folate levels in transgenic rice (Strobbe and Van DerStraeten, 2017). It was reported that overexpression of GTP cyclohydrolase I and aminodeoxychorismate synthase genes enhance the rice folate (vitamin B9) level significantly (Dong *et al.*, 2014). The bioavailability of folates in rice grain is hampered due to its instability and deterioration during long storage. The biofortified high-folate rice grains (150-fold higher than wild rice) were produced by genetic engineering of folate-binding protein, which enhances the stability of folates by binding to them during long storage periods (Blancquaert *et al.*, 2015). Vitamins C, D, and K are generally absent in rice. But knockout of two isoforms of GDP-D-mannose epimerase (OsGME) and GDP-L-galactose phosphorylase (OsGGP) genes produce rice mutants with decreased vitamin C content up to 20% - 30% and 80%, respectively (Ghosh *et al.*,2019).

Gene Editing for Biofortification in Rice

With the help of the effective, practical, and affordable CRISPR-Cas system, important crops can be biofortified to supplement deficiencies in vitamins and minerals. Different genome-edited golden crops, or carotenoid biofortified crops, have been generated to tackle Vit-A deficiency using CRISPR-Cas-based methods. The development of the Golden rice cultivar Kitaake involved the knocking of a 5.2 kb carotenogenesis cassette including the genes for maize PSY and CrtI. The endosperm of this cultivar has 7.9 μg/g dry weight (DW) β-carotene (Dong *et al.*, 2020). Aside from that, recent reports have revealed the use of the CRISPR-Cas9 system for gene editing in other biofortification features. Abe *et al.* (2018) did targeted-mutagenesis in the Fatty acid desaturase gene (*OsFAD2-1*) for producing high oleic/low linoleic in rice bran oil (RBO). Generation of mutations in the phospholipase D (*OsPLDα1*) gene reduced phytic acid content as compared to their wild-type parent (Khan *et al.*, 2019). CRISPR-Cas9 was also used for knocking out *OsAAP6* and *OsAAP10* genes and generation of mutants in *japonica* varieties which reduces the high grain protein content (GPC) (Wang *et al.*, 2020). CRISPR-Cas9-based editing of the *OsNRAMP2* gene caused changes in the remobilization and distribution of Fe and Cd (Chang *et al.*, 2022). Golden rice CRISPR-Cas9 phytoene desaturase CrtI and maize Phytoene synthase (PSY) genes were knocked in Dong *et al.* (2020).

References

Abe, K., Araki, E., Suzuki, Y., Toki, S. and Saika, H. 2018. Production of High Oleic/low Linoleic Rice by Genome Editing. *Plant Physiology Biochem.* **131**, 58–62. doi:10.1016/j.plaphy.2018.04.033

Anai T, Koga M, Tanaka H, Kinoshita T, Rahman SM, Takagi Y. 2003. Improvement of rice (*Oryza sativa* L.) seed oil quality through introduction of a soybean microsomal omega-3 fatty acid desaturase gene. *Plant Cell Rep* **21**(10):988–92. doi:10.1007/s00299-003-0609-6

Blancquaert, D., Van Daele, J., Strobbe, S., Kiekens, F., Storozhenko, S., De Steur, H., Van Der Straeten, D. 2015. Improving folate (Vitamin B9) stability in biofortified rice through metabolic engineering. *Nat. Biotechnol* **33:**1076- 1078.

Blancquaert D, Van daele J, Strobbe S, Kiekens F, Storozhenko S, De Steur H, *et al.* 2015. Improving folate (Vitamin B9) stability in biofortified rice through met-abolic engineering. *Nat Biotechnol* **33:**1076–8. doi:10.1038/nbt.3358

Chang, J.-D., Xie, Y., Zhang, H., Zhang, S., and Zhao, F.J. 2022. The Vacuolar Transporter OsNRAMP2 Mediates Fe Remobilization during Germination and Affects Cd Distribution to Rice Grain. *Plant Soil*, 1–17. doi:10.21203/rs.3.rs- 1206604/v1

Datta K, Baisakh N, Oliva N, Torrizo L, Abrigo E, Tan J, *et al.* 2003. Bioengineered 'golden' Indica rice cultivars with beta-carotene metabolism in the endo-sperm with hygromycin and mannose selection systems. *Plant Biotechnol J* 1:81–90. doi:10.1046/j.1467-7652.2003.00015.x

Dong, W., Cheng, Z., Lei, C., Wang, X., Wang, J., Wang, J., Wan, J. 2014. Overexpression of folate biosynthesis genes in rice (*Oryza sativa* L.) and evaluation of their impact on seed folate content. *Plant Food Hum Nutr*. **69**(4): 379-385.

Fei M, Jin Y, Jin L, Su J, Ruan Y, Wang F, Liu C and Sun C. 2020. Adaptation of Rice to the Nordic Climate Yields Potential for Rice Cultivation at Most Northerly Site and the Organic Production of Low-Arsenic and High-Protein Rice. Front. *Plant Sci*. **11:**329. doi: 10.3389/fpls.2020.00329

Ghosh, S., Datta, K., Datta, S. K. 2019. Rice Vitamins. Rice. https://doi.org/10.1016/B978-0-12-811508-4.00007-1

Garg M, Sharma N, Sharma S, Kapoor P, Kumar A, Chunduri V and Arora P 2018. Biofortified Crops Generated by Breeding, Agronomy, and Transgenic Approaches Are Improving Lives of Millions of People around the World. *Front. Nutr.* **5:**12. doi: 10.3389/fnut.2018.00012

Jiang, S.-Y.Ali Ma, LifenXie and Srinivasan Ramachandran. Improving protein content and quality by over-expressing artificially synthetic fusion proteins with high lysine and threonine constituent in rice plants.*Sci. Rep.* **6**, 34427; doi:10.1038/srep34427 (2016).

Karchi H, Miron D, Benyaacov S, Galili G. 1995. The lysine-dependentstimulation of lysine catabolism in tobacco seed requires calcium andprotein-phosphorylation. The Plant Cell 7, 1963–1970.

Khan, M. S. S., Basnet, R., Islam, S. A., and Shu, Q. (2019). Mutational Analysis of OsPLDα1 Reveals its Involvement in Phytic Acid Biosynthesis in Rice Grains. *J. Agric. Food Chem.* **67**(41), 11436–11443. doi:10.1021/acs.jafc.9b05052

Long, X., Liu, Q., Chan, M., Wang, Q. and Sun, S. S. 2013. Metabolic engineering and profiling of rice with increased lysine. *Plant Biotechnol. J.* **11**: 490–501.

Lee, S. I. *et al.* 2001. Constitutive and seed-specific expression of a maize lysine-feedback in sensitive dihydro dipicolinate synthase gene leads to increased free lysine levels in rice seeds. *Mol. Breeding*. **8:** 75–84.

Liu X, Zhang C, Wang X, Liu Q, Yuan D, Pan G, Sun SSM, Tu J. 2016. Development of high-lysine rice via endosperm-specific expression of a foreign *LYSINE RICH PROTEIN* gene. *BMC Plant Biology* **16:**147, DOI 10.1186/s12870-016-0837-x

Long X, Liu Q, Chan M, Wang Q, Sun SSM. 2013. Metabolicengineering and profiling of rice with increased lysine. *Plant BiotechnologyJournal* **11**: 490–501.

Lee S, Jeon US, Lee SJ, Kim YK, Persson DP, Husted S, *et al.* 2009. Iron fortification of rice seeds through activation of the nicotianamine synthase gene. *Proc Natl Acad Sci USA* **106:** 22014–9. doi:10.1073/pnas.0910950106

Lee S, An G. Over-expression of OsIRT1 leads to increased iron and zinc accumulations in rice. *Plant Cell Environ* (2009) 32:408–16. doi:10.1111/ j.1365-3040.2009.01935.x

Liu Q, Wang Z, Chen X, Cai X, Tang S, Yu H, *et al.* 2003. Stable inheritance of the antisense waxy gene in transgenic rice with reduced amylose level and improved quality. *Transgenic Res.,* **12**(1):71–82. doi:10.1023/ A:1022148824018

Lee JH, Kim IG, Kim HS, Shin KS, Suh SC, Kweon SJ, *et al.* 2010. Development of transgenic rice lines expressing the human lactoferrin gene. *J Plant Biotechnol* **37**(4):556–61. doi:10.5010/JPB.2010.37.4.556

Masuda H, Suzuki M, Morikawa KC, Kobayashi T, Nakanishi H, Takahashi M, *et al.* 2008. Increase in iron and zinc concentrations in rice grains via the intro-duction of barley genes involved in phytosiderophore synthesis. *Rice.,* **1:**100–8. doi:10.1007/s12284-008-9007-6

Ogo Y, Ozawa K, Ishimaru T, Murayama T, Takaiwa F. 2013. Transgenic rice seed synthesizing diverse flavonoids at high levels: a new platform for flavonoid production with associated health benefits. *Plant Biotechnol J* **11:**734–46. doi:10.1111/pbi.12064

Schaeffer, G. W. & Sharpe, F. T. Increased lysine and seed storage protein in rice plants recovered from calli selected with inhibitory levels of lysine plus threonine and S-(2-aminoethyl) cysteine. *Plant Physiol.* **84:** 509–515 (1987).

Strobb, S, Van Der Straeten D 2017. Folate biofortification in food crops. *Curr Opin Biotechnol.,* **44:** 202–211.

Shin YM, Park HJ, Yim SD, Baek NI, Lee CH, An G, *et al.* 2006. Transgenic rice lines expressing maize C1 and R-S regulatory genes produce various flavo-noids in the endosperm. *Plant Biotechnol J* **4:**303–15. doi:10.1111/ j.1467-7652.2006.00182.x

Wang, S., Yang, Y., Guo, M., Zhong, C., Yan, C., and Sun, S. 2020. Targeted Mutagenesis of Amino Acid Transporter Genes for Rice Quality Improvement Using the CRISPR/Cas9 System. *Crop J.* **8**(3):457–464. doi:10.1016/j.cj.2020.02.005

Wong, H. W., Liu, Q. and Sun, S. S. 2015. Biofortification of rice with lysine using endogenous histones. *Plant Mol. Biol.* **87,** 235–248.

Yang QQ, Zhang CQ, Chan MI, *et al.* 2016. Biofortification of rice with theessential amino acid lysine: molecular characterization, nutritional evaluation,and field performance. *Journal of Experimental Botany* **67:** 4258–4296.

Zhu C, Naqvi S, Gomez-Galera S, Pelacho AM, Capell T, Christou P. 2007. Transgenic strategies for the nutritional enhancement of plants. *Trends Plant Sci.,* **12:**548–55. doi:10.1016/j.tplants.2007.09.007

8

Innovation, Constrains and Future Endeavour in Biofortification in Rice in India

Significant Progress and Prospects

It is widely acknowledged that biofortification presents a promising and cost-effective agricultural strategy to enhance the nutritional status of malnourished populations globally. Approaches to biofortification, including crop breeding, targeted genetic modification, and mineral fertilizer application, offer significant potential for combatting mineral deficiencies in human diets. The development of biofortified food crops with enhanced nutrient content such as iron, zinc, selenium, and provitamin A has shown promise in addressing micronutrient deficiencies prevalent in both developing and developed nations. Over the past decade, both national and international initiatives have made substantial contributions toward achieving these goals. These efforts have resulted in the release of biofortified varieties across various crops, including rice, maize, wheat, potatoes, vegetables, and millets. The majority of these varieties have been developed through conventional breeding methods without sacrificing crop yield, making them readily accepted by both growers and consumers.

One of the United Nations' sustainable development goals is the cultivation of nutritionally rich crop varieties with heightened levels of micronutrients such as iron, zinc, calcium, total protein, lysine, tryptophan, anthocyanin, provitamin A, and oleic acid, coupled with reduced levels of anti-nutritional factors. In the past decade, 142 biofortified varieties, spanning rice, wheat, maize, pearl millet, small millet, lentil, chickpea, and other crops, have been developed under the auspices of the Indian Council of Agricultural Research (ICAR). Biofortified varieties typically do not impact ecological conditions, soil, or water requirements differently from traditional varieties. Additionally, they do not incur extra cultivation costs, and their economic output is comparable to traditional produce, leading to their widespread adoption. In India, the scale-up of biofortified varieties has gained momentum, with substantial quantities of breeder seeds being produced and distributed to public and private seed agencies for further multiplication and dissemination to farmers. Over the past six years, approximately 10 million hectares of land have been cultivated with biofortified rice, wheat, pearl millet, mustard, and lentil varieties.

In recognition of the 75th anniversary of the Food and Agriculture Organization (FAO) on October 16, 2020, the Indian Government launched the ambitious POSHAN Abhiyaan aiming to reach over 100 million people to combat stunting, undernutrition, anemia, and low birth weight. As part of this initiative, the government introduced 17 biofortified varieties across eight crops, which offer up to a threefold increase in nutritional value. Notably, the rice variety CR Dhan 315, high in zinc, was among these varieties. The recently unveiled National Nutrition Strategy - 2017 by NITI Aayog, Government of India, aims to address malnutrition through food-based solutions, including the incorporation of biofortified cereals into various government-sponsored programs such as the National Food Security Mission and nutrition intervention programs like the Integrated Child Development Services scheme and the 'Mid-day meal' program. At the international level, the HarvestPlus program leads and coordinates biofortification breeding efforts in staple crops, including rice, to alleviate micronutrient deficiencies in developing countries. Adoption of modern plant breeding and molecular techniques has enabled the enhancement of micronutrients in rice grains, resulting in the release of numerous biofortified rice varieties in several countries, particularly rich in zinc or protein. Bangladesh notably approved the cultivation of β-carotene-enriched golden rice variety BRRI Dhan 29, becoming the first country to do so (Glover *et al.*, 2020).

Constrains and Future Endeavor

Despite significant progress in developing biofortified varieties and understanding the biochemical and molecular mechanisms behind increased nutrient accumulation in grains, several constraints hinder widespread adoption among growers, millers, and consumers. The following challenges outline some future obstacles and potential solutions to bridge the adoption gap. Biofortified Genotypes through Genetic Engineering: While transgenic approaches have been heavily emphasized in the past two decades, regulatory approval processes, especially in countries like India, pose hurdles for biofortified genotypes developed through genetic engineering. Additionally, consumer acceptance remains an issue even for released transgenic lines.

Awareness Generation and Misconception Elimination: Lack of awareness regarding the health benefits of biofortified crops is a major factor slowing down adoption. Educating household heads and involving farmers in demonstration trials and field days can help raise awareness. Addressing misconceptions about low yields of biofortified varieties is crucial, although it's now established that they have comparable yields to traditional varieties.

Acceptance at Consumer Level: In cases where biofortification alters the appearance of grains, such as changing from white to yellow or purple due to increased provitamin-A or anthocyanin, consumers at the village level may hesitate to accept them as food.

Linkages with Agrifood-Processing Industry: Strong connections with the agrifood-processing industry can facilitate the dissemination of biofortified crops. Promotional activities like field demonstrations, TV and radio shows, and live dramas can raise awareness among farmers, industries, and consumers about the existence and benefits of biofortified crops.

Policy Support: Strengthening the seed chain to produce and supply high-quality seeds is crucial for popularizing biofortified varieties. Providing subsidized seeds and other inputs can accelerate the spread of nutritionally improved cultivars among farmers. Ensuring remunerative prices through minimum support price or premium pricing for biofortified grains can incentivize farmers to grow more of these crops. Inclusion of biofortified products in government-sponsored schemes would particularly benefit vulnerable groups such as children, pregnant women, and the elderly. Despite these challenges, biofortified crops offer a promising solution to alleviate micronutrient malnutrition, especially in developing countries like India, and hold a bright future.

Researchable Issues for the Future

Overall, nutrient biofortification of cereal grains by plant breeding and genetic engineering should be mindful of potential perturbations of the plant metabolic network. Non-targeted metabolite profiling of different tissues should be conducted to examine how plants adapt to metabolic modifications in the grains. Along with nutrient stability, nutrient bioavailability as well as the cooking and sensory quality of biofortified cereal grains should also be investigated (Su and Tian 2018).

Desiphering genetic mechanism and underlying genes of grain protein and amino acid content

Rice protein content is governed by polygenes. Many QTLs associated with rice grain protein content have been identified. But only two genes inside those QTL region regulating protein content have been cloned. Majority of these QTLs are with low additive effect and are environment-specific. We understand that the accumulation of rice storage proteins is a complex process involving nitrogen absorption, transport, assimilation, amino acid synthesis, amino acid modification and transport follows by protein synthesis, transport, modification, storage, and degradation. Therefore, for each step, there is enormous scope of analysis of genetic mechanism and identification and cloning of genes inside robust QTL region underlying each genetic mechanism. Pyramiding of those genes will further ensure the stability of grain storage protein and essential amino acids in high yielding introgressed genotypes.

Understanding association of grain protein with cooking and eating quality

An increasing number of studies have shown that grain protein is the most important factor affecting cooking and eating quality after starch. However, more research is needed to fully understand the relationship between grain protein and rice quality, as well as to develop strategies for optimizing it to meet human dietary needs. A series of physical and chemical changes occur in rice during cooking, including water absorption of rice grains, gelatinization of starch, dissolution of starch after endosperm cell breakage, and formation of adhesion layers. An in-depth study in each step, on direct or indirect effect of protein content on cooking, and eating quality will provide a scientific basis for breeders to select and cultivate varieties with superior tastes.

In general, higher N-fertilizer leads to higher grain yield and protein content. On the other hand, the excessive use of nitrogen fertilizer leads to the deterioration of cooking and eating quality of rice. Exploring more nitrogen-use efficient genes and analyzing the mechanism of their influence on protein content in rice will help cultivate nitrogen use efficient varieties. This, in turn, will lead to synergistic improvement in rice yield and quality in optimum use of nitrogen fertilizer (Lou *et al.*, 2023).

Understanding molecular basis of higher content and Increasing bio-availability of micronutrients

Biofortification of rice with increased zinc (Zn) content is an effective strategy for combating Zn deficiency in countries where rice is a staple food. Progress has been made in understanding the molecular mechanisms behind Zn accumulation and the influence of genotype × environment interactions (G × E), resulting in the development and release of Zn-biofortified rice varieties. However, assessing the bioavailability of micronutrients post-consumption is time-consuming and costly, making high-throughput screening of biofortified breeding lines impractical. To address this challenge, suitable in-vitro methodologies are needed.

Phytic acid is a significant anti-nutritional factor that reduces the absorption of iron and zinc in the human body. In the human gut, only a small percentage of iron and zinc in food grains are available for absorption. In rice breeding programs, low-phytic acid mutants (lpa) with reduced phytic acid synthesis are utilized. Knocking out the OsITPK1-6 gene using CRISPR-CAS9 technology results in decreased phytic acid accumulation in rice grains, thereby increasing the availability of micronutrients (Jiang *et al.*, 2019). Furthermore, there is a need to breed for factors such as ascorbic acid and β-carotene, which enhance the bioavailability of iron and zinc.

Easy detection of nutritional quality

Most nutritional traits in rice, including grain protein, amino acids, iron, zinc, and vitamins, are not visibly apparent, making it challenging to persuade farmers and traders about the quality of the produce. Unlike crops like sugarcane and sweet corn, where sugar content can be easily measured using instruments like a Brix meter, there is a lack of handheld and user-friendly equipment for quickly assessing the quality parameters of rice. There is a clear need for the development of portable devices that can provide reliable estimates of rice quality parameters. Such tools would not only promote the adoption of biofortified varieties but also enable government agencies to distinguish them from traditional varieties and provide higher support prices. By empowering farmers and traders with accessible tools to assess rice quality, the adoption of biofortified varieties can be facilitated, leading to improved nutrition outcomes in rice-consuming communities.

Harnessing scope in growing rice based food and beverages industries

In recent times, various formulations of protein, starch, and fibers are being employed to create gluten-free foods. Rice protein, sourced from rice, is a key component of these formulations due to its richness in amino acids, particularly lysine, and its hypoallergenic properties (glutelin instead of gluten). Rice protein offers several advantages, including higher digestibility, a mild taste, and colorless characteristics, making it a preferred choice over other plant-based proteins.

Numerous studies have demonstrated the feasibility of substituting rice protein for allergenic proteins in gluten-free baked goods, catering to individuals with gluten allergies. Additionally, rice protein is utilized in infant formulas to provide nourishment for infants who are allergic to milk protein (Jayprakash *et al.* 2022). Furthermore, rice is increasingly utilized in the production of gluten-free beverages, including beer. All-rice malt beer formulations are gaining popularity in both European and Asian countries. The expanding market for rice-based baby food and beverages presents significant opportunities, particularly in India. By incorporating biofortified rice grains into these products, the nutritional value can be enhanced, contributing to improved health outcomes for consumers.

Conclusion

A significant step forward in advancing biofortified germplasm involves fine mapping of candidate genes associated with various quantitative trait loci (QTLs). With the identification of single nucleotide polymorphisms (SNPs) through next-generation sequencing (NGS), there is a rapid transition in research toward leveraging CRISPR/Cas9 systems for targeted mutagenesis. This approach holds promise for overcoming barriers in breeding improved-quality rice.

By mining the identified genomic/QTL regions, the genes responsible for micronutrient content can be identified, and the development of marker systems would facilitate selection, introgression, and enhancement of micronutrient content in rice (Senguttuvel *et al.* 2023).

Efforts have commenced to integrate grain zinc (Zn) into future rice varieties, ensuring that the Zn trait becomes a fundamental aspect of upcoming varieties. There is considerable potential to apply advanced genomics technologies such as genomic selection and genome editing to accelerate the development of high-Zn varieties. Establishing an efficient rice value chain for Zn-biofortified varieties, alongside quality control and promotion, is crucial for successful adoption and consumption.

Fast-tracking the development of next-generation high-Zn rice varieties with elevated grain-Zn content, combined with stacking multiple nutrients, while maintaining good grain quality and desirable agronomic traits, is imperative. Given the substantial demand for healthier rice from all stakeholders, it is essential to maintain the momentum in developing nutritious rice to meet this demand and achieve nutritional security.

Breeding rice varieties with a comprehensive array of beneficial minerals and vitamins is crucial to develop a holistic biofortified rice product. Priority should be given to efforts aimed at creating rice varieties with high zinc (Zn), iron (Fe), selenium, vitamin A, proteins, amino acids, and other essential nutrients. It would also be intriguing to combine high nutrient content with health-promoting traits such as low glycemic index, antioxidants, and resistant starch. Additionally, there is a need to reduce the levels of harmful elements such as arsenic and cadmium.

The growing demand for rice varieties with improved grain quality and nutrition necessitates the development of a range of rice varieties tailored to different regions and preferences. Nutrition-sensitive agriculture builds upon the foundation laid by the agricultural sector, emphasizing synergies among agricultural, nutritional, environmental, cultural, and economic factors.

Following the Green Revolution of the 1960s, subsequent revolutions focusing on sustainable food production and digital farming have been proposed. The advent of nutrition-sensitive agriculture heralds the fourth Green Revolution, centered on enhancing nutrition security and bridging nutritional gaps in high-yield food crops (Su and Tian 2018).

Given that nutritional traits such as minerals and vitamins do not typically impact grain yield, it's imperative to integrate them into mainstream crop development efforts. High-yielding biofortified crops play a pivotal role in ensuring nutritional security. As Prof. M.S. Swaminathan, the pioneer of the Green Revolution, has emphasized, transitioning from food security to nutrition security requires aligning agriculture, health, and nutrition through collaborative partnerships (Yadava *et al* 2018). To achieve the goal of reaching one billion people by 2030, biofortification efforts must extend beyond initiatives like HarvestPlus. Policymakers need to prioritize the role of agriculture in improving health, while national governments and multilateral institutions should ensure that biofortification is firmly embedded in the nutrition agenda. Public and private sector breeding partners must integrate biofortified traits across their product lines, and food processors and other actors along the value chain should incorporate biofortified crops into their products. Only through concerted efforts spanning the entire value chain can biofortification become standard practice, leading to the realization of reaching one billion people with improved nutrition. (Bouis, HE, Saltzman A. 2017).

References

Bouis, HE, Saltzman A. 2017. Improving nutrition through biofortification: A review of evidence from HarvestPlus, 2003 through 2016. *Global Food Security* **12:** 49–58, http://dx.doi.org/10.1016/j.gfs.2017.01.009.

Glover, D., Kim, S. K., and Stone, G. D. 2020. Golden rice and technology adoption theory: a study of seed choice dynamics among rice growers in the Philippines. *Technol. Soc* **60:** 101227. doi: 10.1016/j.techsoc.2019.101227

Jayaprakash, G.; Bains, A.; Chawla, P.; Fogarasi, M.; Fogarasi, S. 2022. A Narrative Review on Rice Proteins: Current Scenario and Food Industrial Application. *Polymers*, **14:** 3003. https://doi.org/10.3390/ polym14153003.

Jiang, M., Liu, Y., Liu, Y., Tan, Y., Huang, J., and Shu, Q. 2019. Mutation of inositol 1,3,4-trisphosphate 5/6-kinase6 impairs plant growth and phytic acid synthesis in rice. *Plants* **8:** 114. doi: 10.3390/PLANTS8050114

Lou, G., Bhat MA, Tan X, Wang Y., He Y. 2023. Research progress on the relationship between rice protein content and cooking and eating quality and its influencing factors. *Seed Biology*, **2:**16, https://doi.org/10.48130/SeedBio-2023-0016

Senguttuvel P, G P, C J, D SR, CN N, V J, P B, R G, J AK, SV SP, LV SR, AS H, K S, D S, RM S and Govindaraj M 2023. Rice biofortification: breeding and genomic approaches for genetic enhancement of grain zinc and iron contents. *Front. Plant Sci.* **14:**1138408. doi: 10.3389/fpls.2023.1138408

Yadava, DK, Hossain F., Mohapatra T. 2018. Nutritional security through crop biofortification in India: Status & future prospects Indian J Med Res 148, November pp 621-631, DOI: 10.4103/ijmr.IJMR_1893_18

Yu S., Tian L. 2018. Breeding Major Cereal Grains through the Lens of Nutrition Sensitivity. *Molecular Plant* **11:** 23–30

Index